SPEED
Mathematics

SPEED
Mathematics

Secret Skills for Quick Calculation

BILL HANDLEY

WILEY

John Wiley & Sons, Inc.

Dedicated to
Benyomin Goldschmiedt

Published by John Wiley & Sons, Inc., Hoboken, New Jersey
Published simultaneously in Canada

Originally published in Australia under the title *Speed Mathematics: Secrets of Lightning Mental Calculation* by Wrightbooks, an imprint of John Wiley & Sons Australia, Ltd, in 2003.

For general information about our other products and services, please contact our Customer Care Department within the United States at (800) 762-2974, outside the United States at (317) 572-3993 or fax (317) 572-4002.

Wiley also publishes its books in a variety of electronic formats. Some content that appears in print may not be available in electronic books. For more information about Wiley products, visit our web site at www.wiley.com.

ISBN 0-471-46731-6

Printed in the United States of America

10 9 8 7 6 5 4 3 2 1

Contents

Preface

Many people have asked me if my methods are similar to those developed by Jakow Trachtenberg. He inspired millions with his methods and revolutionary approach to mathematics. Trachtenberg's book inspired me when I was a teenager. After reading it I found to my delight that I was capable of making large mental calculations I would not otherwise have believed possible. From his ideas, I developed a love for working, playing and experimenting with numbers. I owe him a lot.

My methods are *not* the same, although there are some areas where our methods meet and overlap. We use the same formula for squaring numbers ending in five. Trachtenberg also taught casting out nines to check answers. Whereas he has a different rule for multiplication by each number from 1 to 12, I use a single formula. Whenever anyone links my methods to Trachtenberg's, I take it as a compliment.

My methods are my own and my approach and style are my own. Any shortcomings in the book are mine.

I am producing a teachers' handbook with explanations of how to teach these methods in the classroom with many handout sheets and problem sheets. Please email me for details.

Bill Handley
bhandley@speedmathematics.com

Introduction

Imagine being able to multiply large numbers in your head—faster than you could tap the numbers into a calculator. Imagine being able to make a "lightning" mental check to see if you have made a mistake. How would your colleagues react if you could calculate square roots—and even cube roots—mentally? Would you gain a reputation for being extremely intelligent? Would your friends and colleagues treat you differently? How about your teachers, lecturers, clients, management?

People equate mathematical ability with intelligence. If you are able to do multiplication, division, squaring and square roots in your head in less time than your friends can retrieve their calculators from their bags, they will believe you have a superior intellect.

I taught a young boy some of the strategies you will learn in *Speed Mathematics* before he had entered first grade and he was treated like a prodigy throughout elementary school and high school.

Engineers familiar with these kinds of strategies gain a reputation for being geniuses because they can give almost instant answers to square root problems. Mentally finding the length of a hypotenuse is child's play using the methods taught in this book.

As these people are perceived as being extremely intelligent, they are treated differently by their friends and family, at school and in the workplace. And because they are *treated* as being more intelligent, they are more inclined to *act* more intelligently.

Why Teach Basic Number Facts and Basic Arithmetic?

Once I was interviewed on a radio program. After my interview, the interviewer spoke with a representative from the mathematics department at a leading Australian university. He said that teaching students to calculate is a waste of time. Why does anyone need to square numbers, multiply numbers, find square roots or divide numbers when we have calculators? Many parents telephoned the network to say his attitude could explain the difficulties their children were having in school.

I have also had discussions with educators about the value of teaching basic number facts. Many say children don't need to know that 5 plus 2 equals 7 or 2 times 3 is 6.

When these comments are made in the classroom I ask the students to take out their calculators. I get them to tap the buttons as I give them a problem. "Two plus three times four equals . . .?"

Some students get 20 as an answer on their calculator. Others get an answer of 14.

Which number is correct? How can calculators give two different answers when you press the same buttons?

This is because there is an order of mathematical functions. You multiply and divide before you add or subtract. Some calculators know this; some don't.

A calculator can't think for you. You must understand what you are doing yourself. If you don't understand mathematics, a calculator is of little help.

Here are some reasons why I believe an understanding of mathematics is not only desirable, but essential for everyone, whether student or otherwise:

⇨ People equate mathematical ability with general intelligence. If you are good at math, you are generally regarded as highly intelligent. High-achieving math students are treated differently by their teachers and colleagues. Teachers have higher expectations of them and they generally perform better—not only at mathematics but in other subject areas as well.

⇨ Learning to work with numbers, especially mastering the mental calculations, will give an appreciation for the properties of numbers.

⇨ Mental calculation improves concentration, develops memory, and enhances the ability to retain several ideas at once. Students learn to work with different concepts simultaneously.

⇨ Mental calculation will enable you to develop a "feel" for numbers. You will be able to better estimate answers.

⇨ Understanding mathematics fosters an ability to think laterally. The strategies taught in *Speed Mathematics* will help you develop an ability to try alternative ways of thinking; you will learn to look for non-traditional methods of problem-solving and calculations.

⇨ Mathematical knowledge boosts your confidence and self-esteem. These methods will give you confidence in your mental faculties, intelligence and problem-solving abilities.

⇨ Checking methods gives immediate feedback to the problem-solver. If you make a mistake, you know immediately and you are able to correct it. If you are right, you have the immediate satisfaction of knowing it. Immediate feedback keeps you motivated.

⇨ Mathematics affects our everyday lives. Whether watching sports or buying groceries, there are many practical uses of mental calculation. We all need to be able to make quick calculations.

Mathematical Mind

Is it true that some people are born with a mathematical mind? Do some people have an advantage over others? And, conversely, are some people at a disadvantage when they have to solve mathematical problems?

The difference between high achievers and low achievers is not the brain they were born with but how they learn to use it. High achievers use better strategies than low achievers.

Speed Mathematics will teach you better strategies. These methods are easier than those you have learned in the past so you will solve problems more quickly and make fewer mistakes.

Imagine there are two students sitting in class and the teacher gives them a math problem. Student A says, "This is hard. The teacher hasn't taught us how to do this. So how am I supposed to work it out? Dumb teacher, dumb school."

Student B says, "This is hard. The teacher hasn't taught us how to do this. So how am I supposed to work it out? He knows what we know and what we can do so we must have been taught enough to work this out for ourselves. Where can I start?"

Which student is more likely to solve the problem? Obviously, it is student B.

What happens the next time the class is given a similar problem? Student A says, "I can't do this. This is like the last problem we had. It's too hard. I am no good at these problems. Why can't they give us something easy?"

Student B says, "This is similar to the last problem. I can solve this. I am good at these kinds of problems. They aren't easy, but I can do them. How do I begin with this problem?"

Both students have commenced a pattern; one of failure, the other of success. Has it anything to do with their intelligence? Perhaps, but not necessarily. They could be of equal intelligence. It has more to do with attitude, and their attitude could depend on what they have been told in the past, as well as their previous successes or failures. It is not enough to tell people to change their attitude. That makes them annoyed. I prefer to tell them they can do better and I will show them how. Let success change their attitude. People's faces light up as they exclaim, "Hey, I can do that!"

Here is my first rule of mathematics:

The easier the method you use to solve a problem, the faster you will solve it with less chance of making a mistake.

The more complicated the method you use, the longer you take to solve a problem and the greater the chance of making an error. People who use better methods are faster at getting the answer and make fewer mistakes, while those who use poor methods are slower at getting the answer and make more mistakes. It doesn't have much to do with intelligence or having a "mathematical brain."

How to Use This Book

Speed Mathematics is written as a non-technical book that anyone can comprehend. By the end of this book, you will understand mathematics as never before, you will marvel that math can be so easy, and you will enjoy mathematics in a way you never thought possible.

Each chapter contains a number of examples. Try them, rather than just read them. You will find that the examples are not difficult. By trying the examples, you will really learn the strategies and principles and you will be genuinely motivated. It is only by trying the examples that you will discover how easy the methods really are.

I encourage you to take your time and practice the examples, both by writing them down and by calculating the answers mentally. By working your way through this book, you will be amazed at your new math skills.

Chapter One

Multiplication: Part One

How well do you know your basic multiplication tables?

How would you like to master your tables up to the 10 times tables in less than 10 minutes? And your tables up to the 20 times tables in less than half an hour? You can, using the methods I explain in this book. I only assume you know the 2 times tables reasonably well, and that you can add and subtract simple numbers.

Multiplying Numbers up to 10

We will begin by learning how to multiply numbers up to 10 × 10. This is how it works:

We'll take 7 × 8 as an example.

Write 7 × 8 = down on a piece of paper and draw a circle below each number to be multiplied.

7 × 8 =
○ ○

Now go to the first number to be multiplied, 7. How many more do you need to make 10? The answer is 3. Write 3 in the circle below the 7. Now go to the 8. What do we write in the circle below the 8? How many more to make 10? The answer is 2. Write 2 in the circle below the 8.

Your work should look like this:

$7 \times 8 =$

③ ②

Now subtract diagonally. Take either one of the circled numbers (3 or 2) away from the number, not directly above, but diagonally above, or crossways. In other words, you either take 3 from 8 or 2 from 7. You only subtract one time, so choose the subtraction you find easier. Either way, the answer is the same, 5. This is the first digit of your answer.

$8 - 3 = 5$ or $7 - 2 = 5$

Now multiply the numbers in the circles. Three times 2 is 6. This is the last digit of your answer. The answer is 56. This is how the completed problem looks.

$7 \times 8 = 56$

③ ②

If you know the 2 times tables reasonably well, you can easily master the tables up to the 10 times table, and beyond. Let's try another example, 8×9.

$8 \times 9 =$

② ①

How many more to make 10? The answer is 2 and 1. We write 2 and 1 in the circles below the numbers. What do we do now? We subtract diagonally.

$8 - 1 = 7$ or $9 - 2 = 7$

Seven is the first digit of your answer. Write it down. Now multiply the two circled numbers.

$2 \times 1 = 2$

Two is the last digit of the answer. The answer is 72.

Isn't that easy? Here are some problems to try by yourself. Instead of writing the answers in the book, you may prefer to write the answers on a piece of paper or in a notebook so that you can do the problems again if you wish.

a) $9 \times 9 =$ e) $8 \times 9 =$

b) $8 \times 8 =$ f) $9 \times 6 =$

c) $7 \times 7 =$ g) $5 \times 9 =$

d) $7 \times 9 =$ h) $8 \times 7 =$

Do all of the problems, even if you know your tables well. This is the basic strategy we will use for almost all of our multiplication.

How did you do? The answers are:

a) 81   b) 64   c) 49   d) 63

e) 72   f) 54   g) 45   h) 56

Isn't this the easiest way to learn your tables?

To Learn or Not to Learn Tables?

Now that you have mastered this method, does it mean you don't have to learn your tables?

The answer is yes and no.

No, you don't have to memorize your tables because you can now, with a little practice, calculate your tables instantly. If you already know your tables then learning this method is a bonus.

The good news is that, if you don't know them, you will learn your tables in record time. After you have calculated $7 \times 8 = 56$ a dozen or more times you will find you remember the answer. In other words, you have learned your tables. Again, this is the easiest method I know to learn your tables, and the most pleasant. And you don't have to worry if you haven't learned them all by heart—you will calculate the answers so quickly that everyone will believe you know them anyway.

Multiplying Numbers Greater Than 10

Does this method work for multiplying large numbers?

It certainly does. Let's try an example:

$96 \times 97 =$

What do we take these numbers up to? How many more to make what? One hundred. So we write 4 under 96 and 3 under 97.

$$96 \times 97 =$$
④ ③

What do we do now? We take away diagonally. 96 minus 3 or 97 minus 4 equals 93. This is the first part of your answer. What do we do next? Multiply the numbers in the circles. Four times 3 equals 12. This is the last part of the answer. The full answer is 9,312.

$$96 \times 97 = 9,312$$
④ ③

Which method is easier, this method or the method you learned in school? This method, definitely.

Remember my first law of mathematics:

The easier the method you use, the faster you do the problem and the less likely you are to make a mistake.

Now, here are some more problems to do by yourself.

a) $96 \times 96 =$ e) $98 \times 94 =$

b) $97 \times 95 =$ f) $97 \times 94 =$

c) $95 \times 95 =$ g) $98 \times 92 =$

d) $98 \times 95 =$ h) $97 \times 93 =$

The answers are:

a) 9,216 b) 9,215 c) 9,025 d) 9,310

e) 9,212 f) 9,118 g) 9,016 h) 9,021

Did you get them all right? If you made a mistake, go back, find where you went wrong and do it again. Because the method is so different from conventional ways of learning tables, it is not uncommon to make mistakes in the beginning.

Racing a Calculator

I have been interviewed on television news programs and documentaries, where they often ask me to compete with a calculator. It usually goes like this. They have a hand holding a calculator in front of the camera and me in the background. Someone from off-screen will call out a problem like 96 times 97. As they call out 96, I immediately take it from 100 and get 4. As they call the second number, 97, I take 4 from it and get an answer of 93. I don't say 93, I say nine thousand, three hundred and . . . I say this with a slow Australian drawl. While I am saying nine thousand, three hundred, I am calculating in my mind, 4 times 3 is 12.

So, with hardly a pause I call, "Nine thousand, three hundred and . . . twelve." Although I don't call myself a "lightning calculator"—many of my students can beat me—I still have no problem calling out the answer before anyone can get the answer on their calculator.

Now do the last exercise again, but this time, do all of the calculations in your head. You will find it is much easier than you imagine. I tell students, you need to do three or four calculations in your head before it really becomes easy; you will find the next time is so much easier than the first. So, try it five times before you give up and say it is too difficult.

Are you excited about what you are doing? Your brain hasn't grown suddenly; you are using it more effectively by using better and easier methods for your calculations.

Chapter Two

Using a Reference Number

We haven't quite reached the end of our explanation for multiplication. The method for multiplication has worked for the problems we have done until now, but, with a slight adjustment, we can make it work for any numbers.

Using 10 as a Reference Number

Let's go back to 7 times 8.

(10)   7 × 8 =

The 10 to the left of the problem is our reference number. It is the number we take our multipliers away from.

Write the reference number to the left of the problem. We then ask ourselves, are the numbers we are multiplying above (higher than) or below (lower than) the reference number? In this case the answer is lower (below) each time. So we put the circles below the multipliers. How much below? Three and 2. We write 3 and 2 in the circles. Seven is 10 minus 3, so we put a minus sign in front of the 3. Eight is 10 minus 2, so we put a minus sign in front of the 2.

(10)   7 × 8 =
    –(3)–(2)

We now work diagonally. Seven minus 2 or 8 minus 3 is 5. We write 5 after the equals sign. Now, multiply the 5 by the reference number, 10. Five times 10 is 50, so write a 0 after the 5. (To multiply any number by ten we affix a zero.) Fifty is our subtotal.

Now multiply the numbers in the circles. Three times 2 is 6. Add this to the subtotal of 50 for the final answer of 56.

Your completed problem should look like this:

$$\text{(10)} \quad 7 \times 8 = 50$$
$$-\text{(3)}-\text{(2)} \quad \underline{+6}$$
$$56 \quad \text{ANSWER}$$

Using 100 as a Reference Number

What was our reference number for 96 × 97 in Chapter One? One hundred, because we asked how many more we needed to make 100. The problem worked out in full would look like this:

$$\text{(100)} \quad 96 \times 97 = 9,300$$
$$-\text{(4)} -\text{(3)} \quad \underline{+12}$$
$$9,312 \quad \text{ANSWER}$$

We need to use this method for multiplying numbers like 6 × 7 and 6 × 6. The method I explained for doing the calculations in your head actually forces you to use this method. Let's multiply 98 by 98 and you will see what I mean.

We take 98 and 98 from 100 and get an answer of 2 and 2. We take 2 from 98 and get an answer of 96. But, we don't say, "Ninety-six." We say, "Nine thousand, six hundred and . . ." Nine thousand, six hundred is the answer we get when we multiply 96 by the reference number of 100. We now multiply the numbers in the circles. Two times 2 is 4, so we can say the full answer of nine thousand six hundred and four.

Do these problems in your head:

a) 96 × 96 =

b) 97 × 97 =

c) 99 × 99 =

d) 95 × 95 =

e) 97 × 98 =

Your answers should be:

a) 9,216   b) 9,409   c) 9,801   d) 9,025   e) 9,506

This is quite impressive because you should now be able to give lightning fast answers to these kinds of problems. You will also be able to multiply numbers below 10 very quickly. For example, if you wanted to calculate 9 × 9, you would immediately "see" 1 and 1 below the nines. One from 9 is 8—you call it 80 (8 times 10). One times 1 is 1. Your answer is 81.

Multiplying Numbers in the Teens

Let us see how we apply this method to multiplying numbers in the teens. We will use 13 times 14 as an example and use 10 as our reference number.

⑩   13 × 14 =

Both 13 and 14 are above the reference number, 10, so we put the circles above the multipliers. How much above? Three and 4; so we write 3 and 4 in the circles above 13 and 14. Thirteen equals 10 plus 3 so we write a plus sign in front of the 3; 14 is 10 plus 4 so we write a plus sign in front of the 4.

+③ +④
⑩   13 × 14 =

As before, we work diagonally. Thirteen plus 4, or 14 plus 3 is 17. Write 17 after the equals sign. Multiply the 17 by the reference number, 10, and get 170. One hundred and seventy is our subtotal, so write 170 after the equals sign.

For the last step, we multiply the numbers in the circles. Three times 4 equals 12. Add 12 to 170 and we get our answer of 182. This is how we write the problem in full:

+③ +④
⑩   13 × 14 = 170
               +12
               182   ANSWER

If the number we are multiplying is above the reference number we put the circle above. If the number is below the reference number we put the circle below.

If the circled numbers are *above* we *add* diagonally, if the numbers are *below* we *subtract* diagonally.

Now, try these problems by yourself.

a) 12 × 15 =   f) 12 × 16 =

b) 13 × 15 =   g) 14 × 14 =

c) 12 × 12 =   h) 15 × 15 =

d) 13 × 13 =   i) 12 × 18 =

e) 12 × 14 =   j) 16 × 14 =

The answers are:

a) 180   b) 195   c) 144   d) 169   e) 168

f) 192   g) 196   h) 225   i) 216   j) 224

If you got any wrong, read through this section again and find what you did wrong, then try them again.

How would you multiply 12 × 21? Let's try it.

$$+②+⑪$$
$$⑩ \quad 12 × 21 =$$

We use a reference number of 10. Both numbers are above 10 so we draw the circles above. Twelve is 2 higher than 10, 21 is 11 more so we write 2 and 11 in the circles. Twenty-one plus 2 is 23, times 10 is 230. Two times 11 is 22, added to 230 makes 252.

This is how your completed problem should look:

$$+②+⑪$$
$$⑩ \quad 12 × 21 = 230$$
$$\underline{+ \; 22}$$
$$252$$

Multiplying Numbers Above 100

Can we use this method for multiplying numbers above 100? Yes, by all means.

To multiply 106 by 104, we would use 100 as our reference number.

(100)   106 × 104 =

The multipliers are higher than or above the reference number, 100, so we draw circles above 106 and 104. How much more than 100? Six and 4. Write 6 and 4 in the circles. They are plus numbers (positive numbers) because 106 is 100 plus 6 and 104 is 100 plus 4.

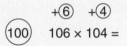

(100)   106 × 104 =

Add crossways. 106 plus 4 is 110. Then, write 110 after the equals sign.

Multiply this number, 110, by the reference number, 100. How do we multiply by 100? By adding two zeros to the end of the number. That makes our subtotal eleven thousand; 11,000.

Now multiply the numbers in the circles. 6 × 4 = 24. Add that to 11,000 to get 11,024.

Our completed calculation looks like this:

$$
\begin{array}{l}
+⑥\quad +④ \\
(100)\quad 106 × 104 = 11,000 \\
\hphantom{(100)\quad 106 × 104 = }\underline{+24} \\
\hphantom{(100)\quad 106 × 104 = }11,024 \quad \text{ANSWER}
\end{array}
$$

Try these for yourself:

a) 102 × 114 =   c) 112 × 112 =

b) 103 × 112 =   d) 102 × 125 =

The answers are:

a) 11,628   b) 11,536   c) 12,544   d) 12,750

With a little practice, you should be able to calculate all of these problems without pencil or paper. That is most impressive.

Solving Problems in Your Head

When we use these strategies, what we visualize or "say" inside our head is very important. It can help us solve problems more easily and more quickly.

Let's calculate for 16×16 and then look at what we would say inside our heads.

Adding diagonally, 16 plus 6 (from the second 16) equals 22, times 10 equals 220. Six times 6 is 36. Add the 30 first, then the 6. Two hundred and twenty plus 30 is 250, plus 6 is 256.

Inside your head you would say, "Sixteen plus six, twenty-two, two twenty. Thirty-six, two fifty-six." With practice, we can leave out half of that. You don't have to give yourself a running commentary on everything you do. You need only say: "Two twenty, two fifty-six."

Practice this. Saying the right things in your head can more than halve the time it takes to do the calculation.

How would you calculate 7×8 in your head? You would "see" 3 and 2 below the 7 and 8. You would take 2 from the 7 (or 3 from the 8) and say "Fifty," multiplying by 10 in the same step. Three times 2 is "Six." All you would say is, "Fifty . . . six."

What about 6×7?

You would "see" 4 and 3 below the 6 and 7. Six minus 3 is 3; you say, "Thirty." Four times 3 is 12, plus 30 is 42. You would just say, "Thirty, forty-two."

It's not as hard as it sounds, is it? And it will become easier the more calculations you do.

When to Use a Reference Number

People ask me, "When should I use a reference number?" The previous example answers this question. When you solve 6 times 7 in your head, you are automatically using a reference number, 10. Your subtotal is 30. You say, "Thirty . . ." Then you calculate 4 times 3 is 12. You wouldn't say "Thirty-twelve." You know you must add the 12 to the 30 to get "Forty-two."

The simple answer is: always use a reference number.

As you become familiar with these strategies you will find you are automatically using the reference number, even if you don't continue to write it down in your calculations.

Combining Methods

Take a look at the following problem:

(100) 92 × 93 =
      –⑧ –⑦

This can still be a difficult calculation if we don't know the answer to 8 × 7. We can draw another pair of circles below the original to multiply 8 × 7. The problem looks like this:

(100) 92 × 93 =
      –⑧ –⑦
      –② –③

Take 8 from 93 by taking 10 and giving back 2. Ninety-three minus 10 equals 83, plus 2 equals 85. Multiply by our reference number, 100, to get a subtotal of 8,500. To multiply 8 × 7, we use the second circled numbers, 2 and 3.

$$7 - 2 = 5 \text{ and } 2 \times 3 = 6$$

The answer is 56. This is how the completed problem would look:

(100) 92 × 93 = 8,500
      –⑧ –⑦ ___56
      –② –③ 8,556   ANSWER

We could also multiply 86 × 87.

(100) 86 × 87 =
      –⑭–⑬

12

86 − 13 = 73

73 × 100 (reference number) = 7,300

+ 182

7,482   ANSWER

We can use the method we have just learned to multiply numbers in the teens.

+④ +③

⑩   14 × 13 = 170

+12

182   ANSWER

You should be able to do this mentally with a little practice.

Try these problems:

a) 92 × 92 =   d) 88 × 85 =

b) 91 × 91 =   e) 86 × 86 =

c) 91 × 92 =   f) 87 × 87 =

The answers are:

a) 8,464   b) 8,281   c) 8,372   d) 7,480

e) 7,396   f) 7,569

Combining the methods taught in this book creates endless possibilities. Experiment for yourself.

Chapter Three

Multiplying Numbers Above and Below the Reference Number

Up until now we have multiplied numbers that were both lower than the reference number or both higher than the reference number. How do we multiply numbers when one number is higher than the reference number and the other is lower than the reference number?

We will see how this works by multiplying 98 × 135. We will use 100 as our reference number:

(100)   98 × 135 =

Ninety-eight is below the reference number, 100, so we put the circle below. How much below? Two, so we write 2 in the circle. One hundred and thirty-five is above 100 so we put the circle above. How much above? Thirty-five, so we write 35 in the circle above.

+(35)
(100)   98 × 135 =
−(2)

One hundred and thirty-five is 100 plus 35 so we put a plus sign in front of the 35. Ninety-eight is 100 minus 2 so we put a minus sign in front of the 2.

We now calculate diagonally. Either 98 plus 35 or 135 minus 2. One hundred and thirty-five minus 2 equals 133. Write 133 down after the equals sign. We now multiply 133 by the reference number, 100. One hundred and thirty-three times 100 is 13,300. (To multiply any number by 100, we simply put two zeros after the number.) This is how your work should look up until now:

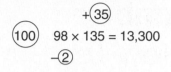

$$(100) \quad 98 \times 135 = 13,300$$

We now multiply the numbers in the circles. Two times 35 equals 70. But that is not really the problem. In fact, we have to multiply 35 by *minus* 2. The answer is *minus* 70. Now your work should look like this:

$$(100) \quad 98 \times 135 = 13,300 - 70 =$$

A Shortcut for Subtraction

Let's take a break from this problem for a moment and look at a shortcut for the subtractions we are doing. What is the easiest way to subtract 70? Let me ask another question. What is the easiest way to take 9 from 56 in your head?

$$56 - 9 =$$

I am sure you got the right answer, but how did you get it? Some would take 6 from 56 to get 50, then take another 3 to make up the 9 they have to take away, and get 47.

Some would take away 10 from 56 and get 46. Then they would add 1 back because they took away 1 too many. This would also give them 47.

Some would do the problem the same way they would using pencil and paper. This way they have to carry and borrow figures in their heads. This is probably the most difficult way to solve the problem. Remember:

> **The easiest way to solve a problem is also the fastest, with the least chance of making a mistake.**

Most people find the easiest way to subtract 9 is to take away 10, then add 1 to the answer. The easiest way to subtract 8 is to take away 10, then add 2 to the answer; and to subtract 7 is to take away 10, then add 3 to the answer. Here are some more "easy" ways:

⇨ What is the easiest way to take 90 from a number?
Take 100 and give back 10.

⇨ What is the easiest way to take 80 from a number?
Take 100 and give back 20.

⇨ What is the easiest way to take 70 from a number?
Take 100 and give back 30.

If we go back to the problem we were working on, how do we take 70 from 13,300? Take away 100 and give back 30. Is this easy? Let's try it. Thirteen thousand, three hundred minus 100? Thirteen thousand two hundred. Plus 30? Thirteen thousand, two hundred and thirty. This is how the completed problem looks:

$$+\!\textcircled{35}$$
$$\textcircled{100} \quad 98 \times 135 = 13,300 - 70 = 13,230 \quad \text{ANSWER}$$
$$-\!\textcircled{2} \qquad\qquad\qquad \textcircled{30}$$

With a little practice you should be able to solve these problems entirely in your head. Practice with the following problems:

a) $98 \times 145 =$ e) $98 \times 146 =$

b) $97 \times 125 =$ f) $9 \times 15 =$

c) $95 \times 120 =$ g) $8 \times 12 =$

d) $96 \times 125 =$ h) $7 \times 12 =$

How did you do? The answers are:

a) 14,210   b) 12,125   c) 11,400   d) 12,000

e) 14,308   f) 135   g) 96   h) 84

Multiplying Numbers in the Circles

The rule for multiplying the numbers in the circles follows.

When both circles are *above* the numbers or both circles are *below* the numbers, we *add* the answer to our subtotal. When one circle is *above* and one circle is *below*, we *subtract*.

Mathematically, we would say: when you multiply two positive (plus) numbers you get a positive (plus) answer. When you multiply two negative (minus) numbers you get a positive (plus) answer. When you multiply a positive (plus) number and a negative (minus) number you get a minus answer.

Would our method work for multiplying 8 × 45?

Let's try it. We choose a reference number of 10. Eight is 2 less than 10 and 45 is 35 more than 10.

$$+\textcircled{35}$$
$$\textcircled{10} \quad 8 \times 45 =$$
$$-\textcircled{2}$$

You either take 2 from 45 or add 35 to 8. Two from 45 is 43, times the reference number, 10, is 430. Minus 2 times 35 is −70. To take 70 from 430 we take 100, which equals 330, then give back 30 for a final answer of 360.

$$+\textcircled{35}$$
$$\textcircled{10} \quad 8 \times 45 = 430 - 70 = 360 \quad \text{ANSWER}$$
$$-\textcircled{2} \qquad\qquad \textcircled{30}$$

Does this replace learning your tables? No, it replaces the method of learning your tables. After you have calculated 7 times 8 equals 56 or 13 times 14 equals 182 a dozen times or more, you stop doing the calculation; you remember the answer. This is much more enjoyable than chanting your tables over and over.

We haven't finished with multiplication yet, but we can take a rest here and practice what we have already learned. If some problems don't seem to work out easily, don't worry; we still have more to cover.

In the next chapter we will look at a simple method for checking our answers.

Chapter Four

Checking Answers: Part One

How would you like to get 100 percent scores on every math test? How would you like to gain a reputation for never making a mistake? Because, if you do make a mistake, I can show you how to find it and correct it, before anyone knows anything about it.

I often tell my students, it is not enough to calculate an answer to a problem in mathematics; you haven't finished until you have checked you have the right answer.

I didn't develop this method of checking answers. Mathematicians have known it for about a thousand years, but it doesn't seem to have been taken seriously by educators in most countries.

When I was young, I used to make a lot of careless mistakes in my calculations. I used to know how to do the problems and I would do everything the right way. But still I got the wrong answer. By forgetting to carry a number, copying down wrong figures and who knows what other mistakes, I would lose points.

My teachers and my parents would tell me to check my work. But the only way I knew how to check my work was to do the problem again. If I got a different answer, when did I make the mistake? Maybe I got it right the first time and made a mistake the second time. So, I would have to solve the problem a third time. If two out of three answers agreed, then that was probably the right answer. But maybe I had made the same mistake twice. So they would tell me to try to solve the problem two different ways. This was good advice. However, they didn't give me time in my math tests to do everything three times. Had someone taught me what I am about to teach you, I could have had a reputation for being a mathematical genius.

I am disappointed that this method was known but nobody taught it. It is called the digit sum method, or casting out nines. This is how it works.

Substitute Numbers

To check a calculation, we use substitute numbers instead of the real numbers we were working with. A substitute on a football or basketball team is somebody who replaces somebody else on the team; they take another person's place. That's what we do with the numbers in our problem. We use substitute numbers instead of the real numbers to check our work.

Let's try an example. Let us say we have just calculated 13 times 14 and got an answer of 182. We want to check our answer.

$$13 \times 14 = 182$$

The first number is 13. We add its digits together to get the substitute:

$$1 + 3 = 4$$

Four becomes our substitute for 13.

The next number we are working with is 14. To find its substitute we add its digits:

$$1 + 4 = 5$$

Five is our substitute for 14.

We now do the original calculation using the substitute numbers instead of the original numbers.

$$4 \times 5 = 20$$

Twenty is a two-digit number so we add its digits together to get our check answer.

$$2 + 0 = 2$$

Two is our check answer.

If we have the right answer in our calculation with the original numbers, the digits in the real answer will add up to the same as our check answer.

We add the digits of the original answer, 182:

$$1 + 8 + 2 = 11$$

Eleven is a two-digit number so we add its digits together to get a one-digit answer:

$$1 + 1 = 2$$

Two is our substitute answer. This is the same as our check answer, so our original answer is correct.

Let's try it again, this time using 13×15:

$$13 \times 15 = 195$$
$$1 + 3 = 4 \text{ (substitute for 13)}$$
$$1 + 5 = 6 \text{ (substitute for 15)}$$
$$4 \times 6 = 24$$

Twenty-four is a two-digit number so we add its digits.

$$2 + 4 = 6$$

Six is our check answer.

Now, to find out if we have the correct answer, we check this against our real answer, 195.

$$1 + 9 + 5 = 15$$

Bring 15 to a one-digit number:

$$1 + 5 = 6$$

Six is what we got for our check answer so we can be confident we didn't make a mistake.

Casting Out Nines

There is another shortcut to this procedure. If we find a 9 anywhere in the calculation, we cross it out. With the previous answer, 195, instead of adding 1 + 9 + 5, we could cross out the 9 and just add 1 + 5 = 6. This makes no difference to the answer, but it saves some work and time. I am in favor of anything that saves time and effort.

What about the answer to the first problem we solved, 182?

We added 1 + 8 + 2 to get 11, then added 1 + 1 to get our final check answer of 2. In 182, we have two digits that add up to 9, the 1 and the 8. Cross them out and you just have the 2 left. No more work at all to do.

Let's try it again to get the idea of how it works.

$$167 \times 346 = 57{,}782$$
$$1 + 6 + 7 = 14$$
$$1 + 4 = 5$$

There were no shortcuts with the first number. Five is our substitute for 167.

$$3 + 4 + 6 =$$

We immediately see that 3 + 6 = 9, so we cross out the 3 and the 6. That just leaves us with 4 as our substitute for 346.

Can we find any nines, or digits adding up to 9 in the answer? Yes, 7 + 2 = 9, so we cross them out. We add the other digits, 5 + 7 + 8 = 20. And 2 + 0 = 2. Two is our substitute answer.

I write the substitute numbers in pencil above or below the actual numbers in the problem. It might look like this:

$$167 \times \cancel{346} = 57{,}\cancel{7}8\cancel{2}$$
$$\quad 5 \qquad 4 \qquad\quad 2$$

Did we get the right answer?

We multiply the substitute numbers, 5 times 4 equals 20, which equals 2 (2 + 0 = 2). This is the same as our substitute answer so we were right.

Let's try one more example:

$$456 \times 831 = 368{,}936$$

We write in our substitute numbers:

~~4~~~~5~~6 × 8~~3~~~~1~~ = ~~3~~~~6~~8,~~9~~~~3~~~~6~~
    6   3   8

That was easy because we cast out (or crossed out) 4 and 5 from the first number, leaving 6; we crossed out 8 and 1 from the second number, leaving 3; and almost every digit was crossed out of the answer, 3 and 6 twice, and a 9.

We now see if the substitutes work out correctly. Six times 3 is 18, which adds up to 9, which also gets crossed out, leaving zero. Our substitute answer is 8, so we got it wrong somewhere.

When we calculate the problem again, we get 378,936.

Did we get it right this time? The 936 cancels out, so we add 3 + 7 + 8 = 18, which adds up to 9 which cancels, leaving 0. This is the same as our check answer, so this time we have it right.

Does casting out nines prove we have the right answer? No, but we can be almost certain (see Chapter Sixteen). For instance, say we got 3,789,360 for our last answer. By mistake we put a zero at the end of our answer. The final zero wouldn't affect our check when casting out nines; we wouldn't know we had made a mistake. But when it showed we had made a mistake, the check definitely proved we had the wrong answer.

Casting out nines is a simple, fast check that will find most mistakes, and should help you achieve 100 percent scores on most of your math tests.

Why Does the Method Work?

Think of a number and multiply it by nine. What are four nines? Thirty-six (36). Add the digits in the answer together (3 + 6) and you get nine.

Let's try another number. Three nines are 27. Add the digits of the answer together, $2 + 7$, and you get 9 again.

Eleven nines are ninety-nine (99). Nine plus 9 equals 18. Wrong answer! No, not yet. Eighteen is a two-digit number so we add its digits together: $1 + 8$. Again, the answer is nine.

If you multiply any number by nine, the sum of the digits in the answer will always add up to nine if you keep adding the digits in the answer until you get a one-digit number. This is an easy way to tell if a number is evenly divisible by nine.

If the digits of any number add up to nine, or a multiple of nine, then the number itself is evenly divisible by nine. That is why, when you multiply any number by nine, or a multiple of nine, the digits of the answer must add up to nine. For instance, say you were checking the following calculation:

$$135 \times 83{,}615 = 11{,}288{,}025$$

Add the digits in the first number:

$$1 + 3 + 5 = 9$$

To check our answer, we don't need to add the digits of the second number, 83,615, because we know 135 has a digit sum of 9. If our answer is correct, it too should have a digit sum of 9.

Let us add the digits in the answer:

$$1 + 1 + 2 + 8 + 8 + 0 + 2 + 5 =$$

Eight plus 1 cancels twice, $2 + 2 + 5 = 9$, so we got it right.

You can have fun playing with this.

If the digits of a number add up to any number other than nine, that number is the remainder you would get after dividing the number by nine.

Let's take 14. One plus 4 is 5. Five is the digit sum of 14. This will be the remainder you would get if you divided by 9. Nine goes into 14 once, with 5 remainder. If you add 3 to the number, you add 3 to the remainder. If you double the number, you double the remainder.

Whatever you do to the number, you do to the remainder, so we can use the remainders as substitutes.

Why do we use nine remainders; couldn't we use the remainders after dividing by, say, 17? Certainly, but there is so much work involved in dividing by 17, the check would be harder than the original problem. We choose nine because of the easy shortcut for finding the remainder.

For more information on why this method works, see Appendix E.

Chapter Five

Multiplication: Part Two

In Chapter One we learned how to multiply numbers using an easy method that makes multiplication fun. This method is easy to use when the numbers are near 10 or 100. But what about multiplying numbers that are around 30 or 60? Can we still use this method? Yes, we certainly can.

We chose reference numbers of 10 and 100 because it is easy to multiply by those numbers. The method will work just as well with other reference numbers, but we must choose numbers that are easy to multiply by.

Multiplication by Factors

It is easy to multiply by 20, as 20 is 2 times 10. And it is simple to multiply by 10 and by 2. This is called multiplication by factors, as 10 and 2 are factors of 20.

$$10 \times 2 = 20$$

Let's try an example:

$$23 \times 24 =$$

Twenty-three and 24 are higher than the reference number, 20, so we put the circles above. How much higher are they? Three and 4. We write these numbers in the circles. We write them above because they are plus numbers (23 = 20 + 3, 24 = 20 + 4).

$$+③ \quad +④$$
$$⑳ \quad 23 \times 24 =$$

We add diagonally as before:

23 + 4 = 27 or,

24 + 3 = 27

Now multiply this answer, 27, by the reference number 20. To do this we multiply by 2, then by 10:

27 × 2 = 54

54 × 10 = 540

(Later, in this chapter, will look at an easy way to multiply numbers like 27 by 2.) The rest is the same as before. We multiply the numbers in the circles and add this to our subtotal:

3 × 4 = 12

540 + 12 = 552

Your work should look like this:

$$+③ \quad +④$$
$$⑳ \quad 23 \times 24 = 27$$
$$540$$
$$\underline{+12}$$
$$552 \quad \text{ANSWER}$$

Checking Our Answers

Let's apply what we learned in Chapter Four and check our answer:

23 × 24 = 552

5   6   12

3

The substitute numbers for 23 and 24 are 5 and 6.

$5 \times 6 = 30$

$3 + 0 = 3$

Three is our check answer.

The digits in our original answer, 552, add up to 3:

$5 + 5 + 2 = 12$

$1 + 2 = 3$

This is the same as our check answer, so we were right.

Let's try another:

$23 \times 31 =$

We put 3 and 11 above 23 and 31 as the multipliers are 3 and 11 above the reference number, 20.

$$+③+⑪$$
$$⑳ \quad 23 \times 31 =$$

Adding diagonally, we get 34:

$31 + 3 = 34$ or $23 + 11 = 34$

We multiply this answer by the reference number, 20. To do this, we multiply 34 by 2, then multiply by 10.

$34 \times 2 = 68$

$68 \times 10 = 680$

This is our subtotal. We now multiply the numbers in the circles, 3 and 11:

$3 \times 11 = 33$

Add this to 680:

$680 + 33 = 713$

The completed calculation will look as follows on the next page.

$$\overset{+\textcircled{3}\ +\textcircled{11}}{\underset{\textcircled{20}}{}}\quad 23 \times 31 = 34$$

$$680$$
$$\underline{+33}$$
$$713 \quad \text{ANSWER}$$

We check by casting out the nines.

$$23 \times 31 = 713$$
$$5 \quad 4 \quad 11$$
$$2$$

Multiply our substitute numbers and then add the digits in the answer:

$$5 \times 4 = 20$$
$$2 + 0 = 2$$

This checks with our substitute answer so we can accept 713 as correct.

Here are some problems to try for yourself. When you have finished them, check your answers by casting out the nines.

a) $21 \times 26 =$ d) $23 \times 27 =$

b) $24 \times 24 =$ e) $21 \times 36 =$

c) $23 \times 23 =$ f) $26 \times 24 =$

You should be able to do all of those problems in your head. It's not difficult with a little practice.

Multiplying Numbers Below 20

How about multiplying numbers below 20? If the numbers (or one of the numbers to be multiplied) are in the high teens, we can use 20 as a reference number.

Let's try an example:

$$19 \times 16 =$$

Using 20 as a reference number we get:

(20) $19 \times 16 =$
  $-(1)$ $-(4)$

Subtract diagonally:

$16 - 1 = 15$ or $19 - 4 = 15$

Multiply by 20:

$15 \times 2 = 30$
$30 \times 10 = 300$

Three hundred is our subtotal.

Now we multiply the numbers in the circles and then add the result to our subtotal:

$1 \times 4 = 4$
$300 + 4 = 304$

Your completed work should look like this:

(20) $19 \times 16 = 15$
  $-(1)$ $-(4)$ 300
                $\underline{+4}$
              304 ANSWER

Let's try the same example using 10 as a reference number.

  $+(9)$ $+(6)$
(10) $19 \times 16 =$

Add diagonally, then multiply by 10 to get a subtotal:

$19 + 6 = 25$
$10 \times 25 = 250$

Multiply the numbers in the circles and add this to our subtotal:

$9 \times 6 = 54$
$250 + 54 = 304$

Your completed work should look like this:

$$+\!\textcircled{9}\ +\!\textcircled{6}$$
$$\textcircled{10}\quad 19 \times 16 = 250$$
$$\underline{+54}$$
$$304 \quad \text{ANSWER}$$

This confirms our first answer.

There isn't much difference between the two reference numbers. It is a matter of personal preference. Simply choose the reference number you find easier to work with.

Numbers Above and Below 20

The third possibility is if one number is above and the other below 20. For example:

$$18 \times 32 =$$
$$+\!\textcircled{12}$$
$$\textcircled{20}\quad 18 \times 32 =$$
$$-\!\textcircled{2}$$

We can either add 18 to 12 or subtract 2 from 32, and then multiply the result by our reference number:

$$32 - 2 = 30$$
$$30 \times 20 = 600$$

We now multiply the numbers in the circles:

$$2 \times 12 = 24$$

It is actually *minus* 2 times 12 so our answer is –24.

$$600 - 24 = 576$$

Your work should look like this:

$$+\!\textcircled{12}$$
$$\textcircled{20}\quad 18 \times 32 = \ 30$$
$$-\!\textcircled{2}\qquad 600 - 24 = 576$$
$$+\!\textcircled{6}$$

(To subtract 24, we subtracted 30 and added 6.)

Let's check the answer by casting out the nines:

$18 \times 32 = 576$

9	5	18
0		0

Zero times 5 is 0, so the answer is correct.

Multiplying Higher Numbers

That takes care of the numbers up to around 30 times 30. What if the numbers are higher? Then we can use 50 as a reference number. It is easy to multiply by 50 because 50 is half of 100, or 100 divided by 2. So, to multiply by 50, we multiply the number by 100, then divide that answer by 2.

Let's try it:

ⓈⓄ 46 × 48 =
    –④ –②

Subtract diagonally:

$46 - 2 = 44$ or $48 - 4 = 44$

Multiply 44 by 100:

$44 \times 100 = 4,400$

To say it in your head, just say, "Forty-four by one hundred is forty-four hundred." Then halve, to multiply by 50, which gives you 2,200.

$4,400 \div 2 = 2,200$

Now multiply the numbers in the circles, and add this result to 2,200:

$4 \times 2 = 8$

$2,200 + 8 = 2,208$

ⓈⓄ 46 × 48 = 4,400
    –④ –②   2,200
              +8
              2,208   ANSWER

Fantastic. That was so easy. Let's try another:

$$53 \times 57 =$$

$$+\textcircled{3} \ +\textcircled{7}$$
$$\textcircled{50} \quad 53 \times 57 =$$

Add diagonally, then multiply the result by the reference number (multiply by 100 and then divide by 2):

$$57 + 3 = 60$$
$$60 \times 100 = 6,000$$
$$6,000 \div 2 = 3,000$$

Multiply the numbers in the circles and add the result to 3,000:

$$3 \times 7 = 21$$
$$3,000 + 21 = 3,021$$

Your work should look like this:

$$+\textcircled{3} \ +\textcircled{7}$$
$$\textcircled{50} \quad 53 \times 57 = 6,000$$

$$3,000$$
$$\underline{+21}$$
$$3,021 \quad \text{ANSWER}$$

Let's try one more:

$$52 \times 63 =$$

$$+\textcircled{2}+\textcircled{13}$$
$$\textcircled{50} \quad 52 \times 63 =$$

Add diagonally and multiply the result by the reference number (multiply by 100 and then divide by 2):

$$63 + 2 = 65$$
$$65 \times 100 = 6,500$$

Then we halve the answer.

If we say "Six thousand five hundred," the answer is easy. "Half of six thousand is three thousand. Half of five hundred is two hundred and fifty. Our sub-total is three thousand, two hundred and fifty."

Now multiply the numbers in the circles:

$2 \times 13 = 26$

Add 26 to our subtotal and we get 3,276. Your work should now look like this:

$$+②+⑬$$
$$⑤⓪ \quad 52 \times 63 = 6,500$$

$$3,250$$
$$\underline{+26}$$
$$3,276 \quad \text{ANSWER}$$

We can check this by casting out the nines.

$$52 \times 63 = 3,276$$
$$7 \quad\quad 0 \quad\quad 0$$

Six plus 3 in 63 adds up to 9, which cancels to leave 0.

In the answer, $3 + 6 = 9$, $2 + 7 = 9$. It all cancels. Seven times zero gives us zero, so the answer is correct.

Here are some problems for you to do. See how many you can do in your head.

a) $46 \times 42 =$ e) $51 \times 55 =$

b) $47 \times 49 =$ f) $54 \times 56 =$

c) $46 \times 47 =$ g) $51 \times 68 =$

d) $44 \times 44 =$ h) $51 \times 72 =$

The answers are:

a) 1,932   b) 2,303   c) 2,162   d) 1,936

e) 2,805   f) 3,024   g) 3,468   h) 3,672

How did you do with those? You should have had no trouble doing all of them in your head. Now check your answers by casting out the nines.

Doubling and Halving Numbers

To use 20 and 50 as reference numbers, we need to be able to double and halve numbers easily.

Sometimes, such as when you must halve a two-digit number and the tens digit is odd, the calculation is not straightforward. For example:

$$78 \div 2 =$$

To halve 78, you might halve 70, then halve 8 and add the answers, but there is an easier method.

$78 = 80 - 2$. Half of $(80 - 2)$ is $(40 - 1)$. That is your answer.

$$40 - 1 = 39$$

To double 38, think of $(40 - 2)$. Double would be $(80 - 4) = 76$.

Try these for yourself:

a) $38 \times 2 =$ d) $68 \times 2 =$

b) $29 \times 2 =$ e) $39 \times 2 =$

c) $59 \times 2 =$ f) $47 \times 2 =$

The answers are:

a) 76   b) 58   c) 118

d) 136   e) 78   f) 94

Now try these:

a) $38 \div 2 =$ e) $34 \div 2 =$

b) $56 \div 2 =$ f) $58 \div 2 =$

c) $78 \div 2 =$ g) $18 \div 2 =$

d) $94 \div 2 =$ h) $76 \div 2 =$

The answers are:

a) 19   b) 28   c) 39   d) 47

e) 17   f) 29   g) 9   h) 38

This strategy can easily be used to multiply or divide larger numbers by 3 or 4. For instance:

$$19 \times 3 = (20 - 1) \times 3 = 60 - 3 = 57$$
$$38 \times 4 = (40 - 2) \times 4 = 160 - 8 = 152.$$

Using 200 and 500 as Reference Numbers

If the numbers you are multiplying are close to either 200 or 500 the calculation is easy because it is easy to use 200 and 500 as reference numbers.

How would you multiply 216 by 216?

If you use 200 as your reference number, the calculation is simple, and easily done in your head.

ⓘ⑥ ⑯
②⓪⓪ $216 \times 216 =$

$216 + 16 = 232$

$232 \times 200 = 46,400$

$(232 \times 2 = 464, \text{ times } 100 = 46,400)$

$16 \times 16 =$

Calculate 16×16 using 10 as a reference:

⑥ ⑥
⑩ $16 \times 16 = 256$

$46,400 + 256 = 46,656$ ANSWER

How about 512×512?

⑫ ⑫
⑤⓪⓪ $512 \times 512 =$

$$512 + 12 = 524$$

524 times 500 is 524 times 1,000 divided by 2.

$524 \times 1,000 = 524,000$ or 524 thousand.

Half of 524 thousand is 262,000.

You could split up the 524 thousand into 500 thousand and 24 thousand. Both are easy to halve mentally.

Half of 500 thousand is 250 thousand. Half of 24 thousand is 12 thousand. 250 thousand plus 12 thousand is 262 thousand.

Now multiply the numbers in the circles:

$$12 \times 12 = 144$$
$$262,000 + 144 = 262,144 \quad \text{ANSWER}$$

Multiplying Lower Numbers

Let's try 6×4:

⑩ $6 \times 4 =$
—④—⑥

We use a reference number of 10. The circles go below because the numbers 6 and 4 are lower than 10. We subtract diagonally:

$$6 - 6 = 0 \ \text{ or } \ 4 - 4 = 0$$

We then multiply the numbers in the circles:

$$4 \times 6 =$$

That was our original problem, (6×4). The method doesn't seem to help. Can we make our method or formula work in this case? We can, but we must use a different reference number. Let's try a reference number of 5. Five is 10 divided by 2, or half of 10. The easy way to multiply by 5 is to multiply by 10 and halve the answer.

+①
⑤ $6 \times 4 =$
    —①

Six is above 5 so we put the circle above. Four is below 5 so we put the circle below. Six is 1 higher than 5 and 4 is 1 lower, so we put 1 in each circle.

We add or subtract diagonally:

$$6 - 1 = 5 \quad \text{or} \quad 4 + 1 = 5$$

We multiply 5 by the reference number, which is also 5.

To do this, we multiply by 10, which gives us 50, and then divide by 2, which gives us 25. Now we multiply the numbers in the circles:

$$1 \times -1 = -1$$

Because the result is a negative number, we subtract it from our sub-total rather than adding it:

$$25 - 1 = 24$$

Thus: ⁺①
  ⑤ $6 \times 4 = 5$
    ⁻① $25 - 1 = 24$ **ANSWER**

This is a long-winded and complicated method for multiplying low numbers, but it shows we can make our method work with a little ingenuity. Actually, these strategies will develop our ability to think laterally. Lateral thinking is very important for mathematicians and also, generally, for succeeding in life.

Let's try some more, even though you probably know your lower tables quite well:

  ⑤ $4 \times 4 =$
  ⁻①⁻①

Subtract diagonally:

$$4 - 1 = 3$$

Multiply your answer by the reference number:

$$3 \times 10 = 30$$
$$30 \div 2 = 15$$

Now multiply the numbers in the circles:

$1 \times 1 = 1$

Add that to our subtotal:

$15 + 1 = 16$

Thus:

⑤ $4 \times 4 = 30$

–①–①   15

+1

16   ANSWER

Try the following:

a) $3 \times 4 =$ d) $3 \times 6 =$

b) $3 \times 3 =$ e) $3 \times 7 =$

c) $6 \times 6 =$ f) $4 \times 7 =$

The answers are:

a) 12   b) 9   c) 36

d) 18   e) 21   f) 28

I'm sure you had no trouble doing those. I don't really think that is the easiest way to learn those tables. I think it is easier to simply remember them. But some people want to learn how to multiply low numbers just to check that the method will work. Others like to know that if they can't remember some of their tables, there is an easy method to calculate the answer. Even if you know your tables for these numbers, it is still fun to play with numbers and experiment.

Multiplication by 5

As we have seen, to multiply by 5 we can multiply by 10 and then halve the answer. Five is half of 10. To multiply 6 by 5, we can multiply 6 by 10, which is 60, and then halve the answer to get 30.

Try these:

  a) $8 \times 5 =$ c) $2 \times 5 =$

  b) $4 \times 5 =$ d) $6 \times 5 =$

The answers are:

 a) 40 b) 20 c) 10 d) 30

This is what we do when the tens digit is odd. Let's try 7×5.

  $7 \times 10 = 70$

If you find it difficult to halve the 70, split it into 60 + 10. Half of 60 + 10 is 30 + 5, which equals 35.

Let's try another:

  $9 \times 5 =$

Ten nines are 90. Ninety splits to 80 + 10. Half of 80 + 10 is 40 + 5, so our answer is 45.

Try these for yourself:

  a) $3 \times 5 =$ c) $9 \times 5 =$

  b) $5 \times 5 =$ d) $7 \times 5 =$

The answers are:

 a) 15 b) 25 c) 45 d) 35

This is an easy method to teach the 5 times multiplication tables. And it works for numbers of any size multiplied by 5.

For example:

  $14 \times 5 =$

  $14 \times 10 = 140$, divided by 2 is 70

Likewise:

  $23 \times 5 =$

  $23 \times 10 = 230$

230 = 220 + 10

Half of 220 + 10 is 110 + 5

110 + 5 = 115

These are lightning mental calculations after just a minute's practice.

Chapter Six

Multiplying Decimals

Numbers are made of digits. 0, 1, 2, 3, 4, 5, 6, 7, 8 and 9 are digits. Digits are like the letters which construct a word. Twenty-three (23) is a two-digit number, made from the digits 2 and 3. The position of the digit in the number determines the digit's place value. For instance, the 2 in the number 23 has a place value of 2 tens, the 3 has a place value of 3 units or ones. Four hundred and thirty-five (435) is a three-digit number. The 4 is the hundreds digit and signifies 4 hundreds, or 400. Three (3) is the tens digit and signifies 3 tens, or 30. Five (5) is the units digit and signifies 5 ones, or simply, 5. When we write a number, the position of the digits is important.

When we write prices, or numbers representing money, we use a decimal point to separate the dollars from the cents. For example, $1.25 represents one dollar, and 25 hundredths of a dollar. The first digit after the decimal point represents tenths of a dollar (ten 10¢ coins make a dollar). The second digit after the decimal point represents hundredths of a dollar (one hundred cents make a dollar).

Multiplying decimals is no more complicated than multiplying any other numbers. This is how we do it.

Let's take an example:

$$1.3 \times 1.4 =$$

(1.3 equals one and three tenths, and 1.4 equals one and four tenths.)

We write down the problem as it is, but ignore the decimal points.

$$+③ \quad +④$$
$$⑩ \quad 1.3 \times 1.4 =$$

Although we write 1.3×1.4, we treat the problem as:

$$13 \times 14 =$$

Ignore the decimal point in the calculation and say, "Thirteen plus four is seventeen, times ten is one hundred and seventy. Four times three is twelve, plus one hundred and seventy is one hundred and eighty-two."

Our work thus far looks like this:

$$+③ \quad +④$$
$$⑩ \quad 1.3 \times 1.4 = 170$$
$$\underline{+12}$$
$$182 \quad \text{ANSWER}$$

However, our problem was 1.3×1.4 and at this point we have calculated 13×14. Our work isn't finished yet. We have to place a decimal point in the answer. To find where we put the decimal point we look at the problem and count the number of digits after the decimal points. There are two digits after the decimal points, the 3 in 1.3, and the 4 in 1.4. Because there are two digits after the decimal points in the problem there must be two digits after the decimal point in the answer. We count two places backwards and put the decimal point between the 1 and the 8, leaving two digits after it.

$$1.82 \quad \text{ANSWER}$$

An easy way to double-check our answer would be to approximate. That means, instead of using the numbers we were given, 1.3×1.4, we round off to 1 and 1.5. 1 times 1.5 is 1.5. The answer should be somewhere between 1 and 2, not 20 or 200. This tells us our decimal point is in the right place.

Let's try another.

$$9.6 \times 97 =$$

We write the problem down as it is, but call the numbers 96 and 97.

⑩⑩   9.6 × 97 =

    –④ –③

96 – 3 = 93

93 × 100 (reference number) = 9,300

4 × 3 = 12

9,300 + 12 = 9,312

Where do we put the decimal point? How many digits follow the decimal point in the problem? One. That's how many digits follow the decimal point in the answer.

931.2   ANSWER

To place the decimal point, we count the total number of digits following the decimal points in both numbers we are multiplying. Make sure we have the same number of digits following the decimal point in the answer. We can double-check our answer by estimating 10 times 90; we know the answer is going to be somewhere near 900, not 9,000 or 90.

If the problem had been 9.6 × 9.7, then the answer would have been 93.12. Knowing this can enable us to take some short cuts that might not be apparent otherwise. We will look at some of these possibilities shortly. In the meantime, try your hand at these problems.

a) 1.3 × 1.3 =   d) 96 × 0.97 =

b) 1.4 × 1.4 =   e) 0.96 × 9.6 =

c) 14 × 0.14 =   f) 13 × 1.5 =

How did you do? The answers are:

a) 1.69   b) 1.96   c) 1.96

d) 93.12   e) 9.216   f) 19.5

What if you had to multiply:

0.13 × 0.14 =

We recall that:

$$13 \times 14 = 182$$

Where do we put the decimal point? How many digits come after the decimal point in the problem? Four, the 1 and 3 in the first number and 1 and 4 in the second. So we count back four digits in the answer. But wait a minute, there are only three digits in the answer. We have to supply the fourth digit. So, we count back three digits, then supply a fourth digit by putting a zero in front of the number.

The answer looks like this:

0.0182   ANSWER

We should also write another zero before the decimal point, because there should always be at least one digit before the decimal point. In this case, we add a zero to make the fourth digit after the decimal point, and place another zero before the decimal point.

Try another to get the idea:

$$0.014 \times 1.4 =$$
$$14 \times 14 = 196$$

Where do we put the decimal point? There are four digits after the decimal point in the problem, 0, 1 and 4 in the first number and 4 in the second. So we must have four digits after the decimal point in the answer. Because there are only three digits in our answer, we supply a zero to make the fourth digit.

Our answer is:

0.0196   ANSWER

Try these for yourself:

a) $23 \times 2.4 =$ c) $0.048 \times 0.48 =$

b) $0.48 \times 4.8 =$ d) $0.0023 \times 0.23 =$

Easy, wasn't it?

Here are the answers:

a) 55.2   b) 2.304   c) 0.02304 d) 0.000529

Understanding this simple principle can help us to solve some problems that appear difficult using our method. They can be adapted to make them easy. Here is an example:

8 × 68 =

What reference number would we use? You could use 10 as the reference number for 8, but 68 is closer to 100. Maybe we could use 50. Our method seems easier when the numbers are close together. So, how do we solve the problem? Why not call the 8, 8.0?

There is no difference between 8 and 8.0. The first number, 8, signifies that the number equals eight. The second number, 8.0, signifies the number equals eight and that it is accurate to one decimal place. The decimal point doesn't change the value.

So, here we go:

(100)   8.0 × 68 =
   —(20)—(32)

Now, the problem is easy. Subtract diagonally.

68 − 20 = 48

Multiply 48 by the reference number (100) 4,800. Multiply the numbers in the circles.

20 × 32 = 640

(To multiply by 20 we multiply by 2 and then by ten, as 2 × 10 = 20.)

4,800 + 640 = 5,440

Thus:

(100)   8.0 × 68 =   4,800
   —(20)—(32)   +640
                  5,440

Now, we have to place the decimal point. How many digits are there after the decimal point in the problem? One, the zero we provided. So we count one digit back in the answer.

544.0   ANSWER

We would normally write the answer as 544.

Try these problems for yourself:

a) $9 \times 83 =$ d) $8 \times 86 =$

b) $9 \times 67 =$ e) $7 \times 89 =$

c) $9 \times 77 =$

The answers are:

a) 747   b) 603   c) 693

d) 688   e) 623

That was easy, wasn't it?

With a little imagination you can use these strategies to solve any multiplication problem.

Chapter Seven

Multiplying Using Two Reference Numbers

Our method for multiplication has worked well for numbers that are close to each other. When the numbers are not close, the method still works but the calculation is more difficult. For instance, what if we wanted to multiply numbers like 13 × 64? Which reference number would we choose? In this chapter, we will look at an easy method to use our strategy with two reference numbers.

It is possible to multiply two numbers that are not close to each other by using two reference numbers. Let me demonstrate a problem, then I will explain how the method works. We will take 8 × 27 as our example. Eight is close to 10, so we will use 10 as our first reference number. Twenty-seven is close to 30, so we will use 30 as our second reference number. From the two reference numbers, we choose the easiest number to multiply by. It is easy to multiply by 10, so we will choose 10. This becomes our base reference number. The second reference number must be a multiple of the base reference number. The number we have chosen, 30, is 3 times the base reference number, 10. Instead of using a circle, I write the two reference numbers to the left of the problem in brackets.

The base reference number is 10. The second reference number is 30, or 3 times 10. We write the reference numbers in parentheses and write the second reference number as a multiple of the first.

$(10 \times 3) \quad 8 \times 27 =$

Both the numbers in the example are lower than their reference numbers, so draw the circles below. Below the 8, which has the base reference number of ten, we draw another circle.

$$(10 \times 3) \quad 8 \times 27 =$$

How much are 8 and 27 lower than their reference numbers (remember the 3 represents 30)? Two and 3. Write 2 and 3 in the circles.

$$(10 \times 3) \quad 8 \times 27 =$$

Now multiply the 2 below the 8 by the multiplication factor, 3, in the parentheses.

$$2 \times 3 = 6$$

Write 6 in the bottom circle below the 2. Then take this bottom circled number, 6, diagonally away from 27.

$$27 - 6 = 21$$

Multiply 21 by the base reference number, 10.

$$21 \times 10 = 210$$

Two hundred and ten, 210, is our subtotal. To get the last part of the answer, multiply the two numbers in the top circles, 2 and 3, to get 6. Add 6 to our subtotal of 210 to get our answer of 216.

$$(10 \times 3) \quad 8 \times 27 = 210$$

$$216 \quad \text{ANSWER}$$

Let's try another:

$$9 \times 48 =$$

Which reference numbers would we choose? Ten and 50. This is how we would write the problem:

(10 × 5)   9 × 48 =

Both numbers are below the reference numbers, so we would draw the circles below. How much below the reference numbers? One and 2. Write 1 and 2 in the circles:

(10 × 5)   9 × 48 =
    –① –②
    –◯

Now multiply the 1 below the 9 by the 5 in the parentheses.

1 × 5 = 5

We write 5 in the circle below the 1. This is how our problem looks now:

(10 × 5)   9 × 48 =
    –① –②
    –⑤

Take 5 from 48:

48 – 5 = 43

Write 43 after the equals sign. Multiply 43 by the base reference number, 10 (write a zero after the 43 to get your answer).

43 × 10 = 430

For the last step, we multiply the numbers in the top circles.

1 × 2 = 2

Add 2 to our subtotal of 430.

430 + 2 = 432

The entire problem looks like this:

(10 × 5)   9 × 48 = 430
    –① –②   +2
    –⑤   432   ANSWER

The calculation part of the problem is easy. The only difficulty you may have is remembering what to do next.

If the numbers are greater than the reference numbers, we do the calculation like this. We will take 13 × 42 as our example:

$$+\!\!\textcircled{12}$$
$$+\!\!\textcircled{3} \quad +\!\!\textcircled{2}$$
$$(10 \times 4) \quad 13 \times 42 =$$

The base reference number is 10. The second reference number is 40, or 10 times 4. We try to choose reference numbers that are both greater than or both less than the numbers being multiplied. Both numbers in this example are greater, so we draw the circles above. Thirteen has the base reference number of 10 so we draw two circles above the 13. How much more than the reference numbers are 13 and 42? Three and 2. Write 3 and 2 in the circles. Multiply the 3 above 13 by the multiplication factor in the parentheses, 4.

$$3 \times 4 = 12$$

Write 12 in the top circle above 13. Now add diagonally.

$$42 + 12 = 54$$

Fifty-four times our base number of 10 is 540. This is our subtotal. Now multiply the numbers in the first circles.

$$3 \times 2 = 6$$

Add 6 to 540 for our final answer of 546. This is how the finished problem looks:

$$+\!\!\textcircled{12}$$
$$+\!\!\textcircled{3} \quad +\!\!\textcircled{2}$$
$$(10 \times 4) \quad 13 \times 42 = 540$$
$$\underline{+6}$$
$$546 \quad \text{ANSWER}$$

The base reference number does not have to be 10. To multiply 23 × 87 you would use 20 as your base reference number and 80 (20 × 4) as your second.

Let's try it:

(20×4) $23 \times 87 =$

Both numbers are higher than the reference numbers, 20 and 80, so we draw the circles above. How much higher? Three and 7. Write 3 and 7 in the circles.

$+\bigcirc$
$+\!\!\textcircled{3}\ +\!\!\textcircled{7}$
(20×4) $23 \times 87 =$

We multiply the 3 above the 23 by the multiplication factor in the parentheses, 4.

$3 \times 4 = 12$

Write 12 above the 3. Your work will look like this:

$+\!\!\textcircled{12}$
$+\!\!\textcircled{3}\ +\!\!\textcircled{7}$
(20×4) $23 \times 87 =$

Then add the 12 to the 87.

$87 + 12 = 99$

We multiply 99 by the base reference number, 20.

$99 \times 20 = 1,980$

(We multiply 99 by 2 and then by 10. Ninety-nine is 100 minus 1. Two times 100 minus 1 is 200 minus 2. And 200 minus 2 equals 198. Now multiply 198 by 10 to get our answer for 99×20.)

Now multiply the numbers in the original circles.

$3 \times 7 = 21$
$1,980 + 21 = 2,001$

The finished problem looks like this:

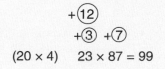

$+\!\!\textcircled{12}$
$+\!\!\textcircled{3}\ +\!\!\textcircled{7}$
(20×4) $23 \times 87 = 99$

$$1,980$$
$$\underline{+21}$$
$$2,001 \quad \text{ANSWER}$$

Here are some more to try by yourself:

a) $14 \times 61 =$ c) $8 \times 136 =$

b) $96 \times 389 =$

To calculate 8×136, problem c), you would use 10 and 140 (10×14) as reference numbers. The answers are:

a) 854   b) 37,344   c) 1,088

Let's calculate b) and c) together:

b) $96 \times 389 =$

We use 100 and 400 as our reference numbers.

$(100 \times 4) \quad 96 \times 389 =$
$\quad\quad\quad\quad -④ \ -⑪$

Multiply the 4 below the 96 by the multiplication factor, 4.

$4 \times 4 = 16$

Now write 16 below the 4 below the 96. Our work looks like this:

$(100 \times 4) \quad 96 \times 389 =$
$\quad\quad\quad\quad -④ \ -⑪$
$\quad\quad -⑯$

We now subtract 16 from 389 and get 373. Next, we multiply the 373 by the base reference number, 100, to get an answer of 37,300.

$(100 \times 4) \quad 96 \times 389 = 37,300$
$\quad\quad\quad\quad -④ \ -⑪$
$\quad\quad -⑯$

We now multiply 4 by 11 in the circles to get 44. Add 44 to 37,300 to get an answer of 37,344.

The completed problem looks like this:

$$(100 \times 4) \quad 96 \times 389 = 37{,}300$$

-④ -⑪ _+44_

-⑯   37,344   ANSWER

Let's try c):

$$8 \times 136 =$$

We use 10 and 140 (10 × 14) as our reference numbers.

$$(10 \times 14) \quad 8 \times 136 =$$

-②   -④

We multiply the 2 below the 8 by the 14 in the parentheses.

$$2 \times 14 = 28$$

Write 28 below the 2 below the 8. Now subtract 28 from 136 (take 30 and add 2) to get 108. We multiply 108 by the base reference number, 10, to get 1,080. Our work looks like this so far:

$$(10 \times 14) \quad 8 \times 136 = 1{,}080$$

-②   -④

-㉘

Now we multiply the numbers in the original circles.

$$2 \times 4 = 8$$

Add 8 to 1,080 to get an answer of 1,088.

$$(10 \times 14) \quad 8 \times 136 = 1{,}080$$

-②   -④   _+8_

-㉘   1,088   ANSWER

Using Factors Expressed as a Division

To multiply 96 × 47, you could use reference numbers of 100 and 50: (50 × 2) or (100 ÷ 2). Here, (100 ÷ 2) would be easier because 100 then becomes your base reference number. It is easier to multiply by

100 than it is by 50. Note, when writing the multiplication problem, write the number first which has the base reference number.

For example:

$$96 \times 47 =$$

Use 100 and 50 as reference numbers.

$$(100 \div 2) \quad 96 \times 47 =$$
$$-\textcircled{4} \ -\textcircled{3}$$

Divide the 4 below 96 by the 2 in the parentheses.

$$4 \div 2 = 2$$

Now we write this 2 in the second circle below 96.

$$(100 \div 2) \quad 96 \times 47 = 4,500$$
$$-\textcircled{4} \ -\textcircled{3}$$
$$-\textcircled{2}$$

Now, subtract 2 from 47 and multiply the answer, 45, by the base reference number, 100. This gives our subtotal, 4,500.

$$(100 \div 2) \quad 96 \times 47 = 4,500$$
$$-\textcircled{4} \ -\textcircled{3} \quad \underline{+12}$$
$$-\textcircled{2} \qquad 4,512 \quad \text{ANSWER}$$

Next, multiply the first two circles ($-4 \times -3 = 12$), and add this to our subtotal. This gives our answer of 4,512.

$$(100 \div 2) \quad 96 \times 47 = 4,500$$
$$-\textcircled{4} \ -\textcircled{3} \quad \underline{+12}$$
$$-\textcircled{2} \qquad 4,512 \quad \text{ANSWER}$$

If you were multiplying 96 × 23 you could use 100 as your base reference number and 25 (100 ÷ 4) as your second reference number. You would write it like this:

$$(100 \div 4) \quad 96 \times 23 =$$
$$-\textcircled{4} \ -\textcircled{2}$$

Ninety-six is 4 lower than 100 and 23 is 2 lower than 25. We now divide the 4 below 96 by the 4 in the parentheses. Four divided by 4 is 1. Write this number below the 4 below the 96.

$$(100 \div 4) \quad 96 \times 23 =$$
$$-④ \quad -②$$
$$-①$$

Subtract 1 from 23 to get 22. Multiply this 22 by the base reference number, 100, to get 2,200.

$$(100 \div 4) \quad 96 \times 23 = 2,200$$
$$-④ \quad -②$$
$$-①$$

Multiply the numbers in the original circles.

$$4 \times 2 = 8$$

Add this to 2,200 to get our answer of 2,208.

$$(100 \div 4) \quad 96 \times 23 = 2,200$$
$$-④ \quad -② \quad \underline{+8}$$
$$-① \qquad 2,208 \quad \text{ANSWER}$$

What if we had multiplied 97 by 23? Would our strategy still work? Let's try it:

$$(100 \div 4) \quad 97 \times 23 =$$
$$-③ \quad -②$$
$$-③④$$

Three divided by 4 is 3 over 4 , or ¾. Take ¾ from 23. (Take 1 and give back a ¼.)

$$23 - ¾ = 22¼$$

One quarter expressed as a decimal is 0.25 (¼ of 100 equals 25). Hence:

$$22¼ \times 100 = 2,225$$

Multiply the numbers in the circles.

$$3 \times 2 = 6$$

$$2,225 + 6 = 2,231$$

$(100 \div 4)$ $97 \times 23 = 2,225$

            $-③$ $-②$ $+6$

            $-③⁄₄$ $2,231$ ANSWER

So our method works well for such problems.

How about 88×343? We can use base numbers of 100 and 350.

$(100 \times 3½)$ $88 \times 343 =$

            $-⑫$ $-⑦$

            $-㊷$

To find the answer to $3½ \times 12$, you multiply 12 by 3, which is 36, and then add half of 12, which is 6, to get 42.

$$343 - 42 = 301$$

$301 \times 100 \text{ (base reference number)} = 30,100$

$$12 \times 7 = 84$$

$$30,100 + 84 = 30,184$$

$(100 \times 3½)$ $88 \times 343 = 30,100$

            $-⑫$ $-⑦$ $+84$

            $-㊷$ $30,184$ ANSWER

Why Does This Method Work?

I won't give a full explanation, but this may help. Let's take an example of 8×17. We could double the 8 to make 16, multiply it by 17, and then halve the answer to get the correct answer for the original problem. This is a hard way to go about it, but it will illustrate why the method using two reference numbers works. We will use a reference number of 20.

⑳ $16 \times 17 =$

        $-④$ $-③$

Subtract 4 from 17 and you get 13. Multiply the 13 by the reference number, 20, to get 260. Now multiply the numbers in the circles.

$4 \times 3 = 12$

Add 12 to the previous answer of 260 for a final answer of 272. But we multiplied by 16 instead of 8, so we have doubled the answer. Two hundred and seventy-two divided by 2 gives us our answer for 8 times 17 of 136.

⑳ $16 \times 17 = 13$
   −④ −③   260
                   +12
                   272   ANSWER

Half of 272 is 136. Thus:

$8 \times 17 = 136$

Now, we doubled our multiplier at the beginning and then halved the answer at the end. These two calculations cancel. We can leave out a considerable portion of the calculation to get the correct answer. Let's see how it works when we use the two reference number method.

(10×2) $8 \times 17 = 130$
   −② −③   +6
   −④   136   ANSWER

Notice that we subtracted 4 from 17 in the second calculation, the same as we did in the first. We got an answer of 13, which we multiplied by ten. In the first calculation we doubled the 13 before multiplying by ten, then we halved the answer at the end. The second time we ended by multiplying the original circled numbers, 2 and 3, to get an answer of 6, half the answer of 12 we got in the first calculation.

You can use any combination of reference numbers. The general rules are:

⇨   First, make the base reference number an easy number to multiply by, e.g., 10, 20, 50.

⇨ The second reference number must be a multiple of the base reference number, e.g., double the base reference number, three times, ten times, fourteen times.

Play with these strategies. There is no limit to what you can do to make math calculations easier. And each time you use these methods you develop your mathematical skills.

Chapter Eight

Addition

Most of us find addition easier than subtraction. We will learn in this chapter how to make addition even easier.

How would you add 43 plus 9 in your head?

The easy way would be to add 10, (53), and take away 1. The answer is 52.

It is easy to add 10 to any number; 36 plus 10 is 46; 34 plus 10 is 44, etc. Simply increase the tens digit by 1 each time you add 10 to a number. (See Chapter Six for an explanation of the breakdown of numbers into digits.)

Here is a basic rule for mental addition:

> **To add 9, add 10 and subtract 1; to add 8, add 10 and subtract 2; to add 7, add 10 and subtract 3, and so on.**

If you wanted to add 47, you would add 50 and subtract 3. To add 196, add 200 and subtract 4. This makes it easy to hold numbers in your head. To add 38 to a number, add 40 and subtract 2. To add 288 to a number, add 300 and subtract 12.

Try these quickly in your head. Call out the answers. For 34 + 9, don't call out, "Forty-four, forty-three." Make the adjustment in your head while you call out the answer. Just say, "Forty-three." Try the following examples. Help is given with two of them.

a) 56 + 8 =

    -②

b) 38 + 9 =

c) 76 + 9 =

d) 65 + 9 =

    -①

e) 47 + 8 =

f) 26 + 7 =

The answers are:

a) 64   b) 47   c) 85

d) 74   e) 55   f) 33

Two-Digit Mental Addition

How would you add 38? To add 38, you would add 40 and subtract 2.

How about 57? You would add 60 and subtract 3.

How would you add 86? To add 86, you would add 100 and subtract 14.

There is a simple principle for mental addition.

> **If the units digit is high, round off to the next ten and then subtract the difference. If the units digit is low, add the tens, then the units.**

With two-digit mental addition you add the tens digit of each number first, then the units. If the units digit is high, round off the number upwards and then subtract the difference. For instance, if you are adding 47, add 50, and then subtract 3.

To add 35, 67 and 43 together you would begin with 35, add 70 to get 105, subtract 3 to get 102, add 40 to get 142 then the 3 to get your answer of 145.

With a little practice, you will be amazed at how you can keep the numbers in your head. Try these for yourself:

a) 34 + 48 =

b) 62 + 26 =

c) 82 + 39 =

d) 27 + 31 =

e) 33 + 44 =

f) 84 + 76 =

g) 44 + 37 =

The answers are:

  a) 82   b) 88   c) 121   d) 58

  e) 77   f) 160   g) 81

For the last example you could have seen that 37 is 3 less than 40, so you could add 40 and then subtract 3; or you could have taken the 3 from 44 before you added to get 41, and added 40 to get your answer of 81. When you actually calculate the problems mentally you find it is not so difficult and you will recognize short cuts as you go.

Adding Three-Digit Numbers

We use the same method for adding three-digit numbers.

To add 355, 752 and 694 together you would say in your head, "Three fifty-five, add seven hundred, (ten fifty-five), add fifty, (eleven hundred and five), plus two, (eleven oh seven), plus seven hundred less six, eighteen hundred and one." Or, you may prefer to add from left to right; adding the hundreds first, then the tens and then the units.

With a little practice, you will find such problems very easy.

Here are some to try for yourself:

  a) 359 + 523 =   c) 456 + 298 =

  b) 123 + 458 =   d) 345 + 288 =

The answers are:

  a) 882   b) 581   c) 754   d) 633

For a), you would round off 359 to 360. Five hundred and twenty-three plus 300 is 823; plus 60 is 883, minus 1 is 882. Or, you could say, 360 plus 500 is 860, plus 23 is 883, minus 1 is 882. Either way is easy.

You should have found c) very easy. You would round 298 to 300.

    456 + 300 = 756

    756 − 2 = 754

Mental addition is easier than the effort of finding a pen and paper or retrieving a calculator from your bag.

Adding Money

If we had to add $4.95, $6.95 and $13.95, we would add $5 + $7 + $14 and subtract 15¢ from the total. Add the tens first, then the units to get 12, 22, 26.

$$5 + 7 = 12$$
$$12 + 14 = 26$$

Our answer is $26.00. We then subtract 15¢ to reach an answer of $25.85.

If we had to add:

$$
\begin{array}{r}
495 \\
695 \\
+ \ 1{,}395 \\
\hline
\end{array}
$$

then the answer might not be so obvious. Yet it is the same problem. We might naturally use that method for adding sums of money (not many do) but fail to see that it is precisely the same problem without the dollar sign and decimals.

If you find one value repeated many times in the calculation, multiply that value by the number of times it appears. Let's say you had to add the following numbers:

$$
\begin{array}{r}
119.95 \\
59.95 \\
119.95 \\
119.95 \\
14.95 \\
+ \ 119.95 \\
\hline
\end{array}
$$

These may be, for instance, the prices of materials sold by a local supplier. How would you add them?

Firstly, you would round off each price (number) to an easy number. $119.95 would be rounded off to $120. $59.95 would be rounded up to $60, and $14.95 would be rounded up to $15. You would adjust for each 5¢ when you have finished.

Secondly, because the price $119.95 appears four times, you would multiply it by 4, and then add the other values. $120 times 4 equals $480. Then add $60 and $15. 480 plus 60 is 540, plus 15 gives us 555.

Now adjust for the 5¢ we previously added to round each price off. There were six prices to which we added 5¢. We multiply 5¢ by 6 items to get a total of 30¢ we added. Subtract this to correct our answer.

$555 minus 30¢ is $554.70. (Take $1 and give back 70¢.)

Adding Larger Numbers

Here is an example of a mental addition with larger numbers:

$$\begin{array}{r} 8,461 \\ + \underline{5,678} \end{array}$$

We begin with the thousands column.

Eight plus 5 is 13. Because we are working in thousands, our answer is 13 thousand. We observe that the numbers in the hundreds column conveniently add to 10, so that gives us another thousand. Our running total is now 14 thousand.

Add the 61 from the top figure.

Fourteen thousand and sixty-one.

Then add 78.

I would add 80 and subtract 2. To add 80, I add 100 and subtract 20. So we have to add 100, subtract 20, then subtract another 2.

Fourteen thousand and sixty-one plus 100 equals 14,161. Less 20 is 14,141. Less 2 is 14,139.

Another method is to begin with the first number, 8,461, and add the second number in parts, beginning with the thousands, then the hundreds, the tens and the units.

You would say, "Eight thousand, four hundred and sixty-one plus five thousand, thirteen thousand, four hundred and sixty-one, plus six hundred is fourteen thousand and sixty-one, plus seventy-eight." Add the 78 as above.

Try these for yourself, adding from left to right:

a) 3,456
+ 3,914

d) 2,750
+ 5,139

b) 9,785
+ 1,641

e) 2,156
+ 2,498

c) 4,184
+ 1,234

Answers:

a) 7,370   b) 11,426   c) 5,418

d) 7,889   e) 4,654

Let's say we have to add the following numbers:

6
8
+ 4

The easy way to add the numbers would be to add:

6 + 4 = 10, plus 8 = 18

Most people would find that easier than 6 + 8 + 4 = 18. (Six plus 8 equals 14, plus 4 equals 18.)

So, an easy rule is:

When adding a column of numbers, add pairs of digits to make tens first, then add the other digits.

Also, add a digit to make the total up to the next multiple of ten. That is, if you have reached, say, 27 in your addition, and the next two numbers to add are 8 and 3, add the 3 before the 8 to make 30, and then add 8 to make 38. Using our methods of multiplication will imprint the combinations of numbers that add to 10 in your mind, and this should become automatic.

Checking Addition by Casting Out Nines

Just as we cast out nines to check our answers for multiplication, so we can use the strategy for checking our addition and subtraction.

Here is an example:

```
  1 2 3 4 5
  6 7 8 9 0
  4 2 7 3 5
+ 2 1 8 6 5
```

We add the numbers to get our total of 144,835.

Is our answer correct?

Let's check by casting out nines, or working with our substitute numbers.

```
  1 2 3 4 5            6
  6 7 8 9 0     21     3
  4 2 7 3 5            3
+ 2 1 8 6 5     13     4
  1 4 4 8 3 5          7
```

Our substitutes are 6, 3, 3 and 4. The first 6 and 3 cancel to leave us with just 3 and 4 to add. 3 + 4 = 7. Seven is our check or substitute answer.

The real answer should add to 7. Let's check and see.

$$1 + 4 + 4 + 8 + 3 + 5 = 7$$

Our answer is correct.

If the above figures were amounts of money with a decimal it would make no difference. You can use this method to check almost all of your additions, subtractions, multiplications and divisions.

Try these for yourself. Are all of these answers right? Check them by casting out the nines. If there is a mistake, correct it and check your answer by casting out the nines.

a) 12,345	b) 25,137	c) 58,235
67,890	15,463	21,704
2,531	51,684	+ 97,105
+ 72,406	+ 25,170	177,144
155,172	17,454	

Chapter Nine

Subtraction

Most people find subtraction more difficult than addition. It need not be so. We will learn some strategies in this chapter that will make subtraction easy.

Firstly, we will deal with mental subtraction.

To subtract mentally, try and round off the number you are subtracting and then correct the answer.

To subtract 9, take 10 and add 1; to subtract 8, take 10 and add 2; to subtract 7, take 10 and add 3. For instance:

$$56 - 9 =$$
①

To take 9 from 56 in your head, the easiest and fastest method is to subtract 10, (46) and add 1, (47).

To take 8 from 47, take 10, (37) and add 2 (39).

To subtract 38 from 54, take 40, (14) and add 2 to get your answer of 16.

Written down, the problem should look like this:

$$54 - 38 = 16$$
②

Fifty-four minus 40 is 14, plus 2 (in the circle) makes 16.

To subtract a number near 100, take 100 and add the remainder. For instance, to subtract 87 from a number, take 100 and add 13, as 100 is 13 more than you wanted to subtract.

$$436 - 87 =$$
$$\textcircled{13}$$

Take 100 to get 336. Add 13 (add 10 and then 3) to get 349. Easy.

Subtracting One Number Below a Hundreds Value From Another Which Is Just Above the Same Hundreds Number

If the number you are subtracting is below 100 and the number you are subtracting from is above 100 and below 200, here is an easy method to calculate in your head.

For example:

$$134 - 76 =$$
$$\textcircled{24}$$

Seventy-six is 24 lower than 100. One hundred and thirty-four is 34 higher than 100. Add 24 to 34 for an easy answer of 58.

Let's try another:

$$152 - 88 =$$
$$\textcircled{12}$$
$$12 + 52 = 64 \quad \text{ANSWER}$$

Try these for yourself:

a) $142 - 88 =$ d) $114 - 80 =$

b) $164 - 75 =$ e) $112 - 85 =$

c) $123 - 70 =$ f) $136 - 57 =$

How did you do?

They are easy if you know how.

Here are the answers:

a) 54   b) 89   c) 53

d) 34   e) 27   f) 79

If you made any mistakes, go back and re-read the explanation then try them again.

The same principle applies for numbers above and below ten. For instance:

$$13 - 6 =$$

$$+ \textcircled{4}$$

$$3 + 4 = 7 \quad \text{ANSWER}$$

Try these for yourself:

a) 12 − 7 =   c) 13 − 9 =

b) 15 − 8 =   d) 14 − 8 =

The answers are:

a) 5   b) 7   c) 4   d) 6

Also, the strategy works for any three-digit subtraction.

$$461 - 275 =$$

$$\textcircled{25}$$

$$461 - 300 = 161$$

$$161 + 25 = 186$$

We do one easy subtraction; the rest is addition.

Let's try another:

$$834 - 286 =$$

$$\textcircled{14}$$

$$834 - 300 = 534$$

$$534 + 14 = 548$$

Say 14 to make 300, plus 500 to make 800 (514), plus 34 is 548.

68

To add the 34, you can add 30, then 4.

This is an easy method for mental subtraction. There is no carrying or borrowing, and it is not difficult to keep track of the numbers in your head.

Try these for yourself:

a) 541 − 87 =   c) 725 − 375 =

b) 263 − 198 =   d) 429 − 168 =

The answers are:

a) 454   b) 65   c) 350   d) 261

For the last problem you could round off 429 to 430 and adjust for the 1 at the end of the calculation.

Written Subtraction

Here is the method I was taught in third grade for written subtraction. If you have practiced the multiplication method you will find this easy.

Easy subtraction uses either of two carrying and borrowing methods. You should recognize one or even both methods.

The difference between standard subtraction and easy subtraction is minor, but important. I will explain easy subtraction with two methods of carrying and borrowing. Use the method you are used to or that you find easier.

Subtraction: Method One

Here is a typical subtraction: 7,254 − 3,897 =

$$
\begin{array}{r}
6\ ^{1}1\ ^{1}4 \\
\not{7}\ \not{2}\ \not{5}\ ^{1}4 \\
-\ 3\ 8\ 9\ 7 \\
\hline
3\ 3\ 5\ 7
\end{array}
$$

Subtract 7 from 4. You can't, so you borrow 1 from the tens column. Cross out the 5 and write 4. Now, here is the difference. You don't say

7 from 14, you say 7 from 10 = 3, then add the number above (4) to get 7, the first digit of the answer.

With this method, you never subtract from any number greater than 10. The rest is addition. Nine from 4 won't go, so borrow again. Nine from 10 is 1, plus 4 is 5, the next digit of the answer.

Eight from 1 won't go, so borrow again. Eight from 10 is 2, plus 1 is 3, the next digit of the answer.

Three from 6 is 3, the final digit of the answer.

Subtraction: Method Two

$$7 \; {}^{1}2 \; {}^{1}5 \; {}^{1}4$$
$$-\,{}_{1}3 \;{}_{1}8 \;{}_{1}9 \; 7$$
$$3 \; 3 \; 5 \; 7$$

Subtract 7 from 4. You can't, so you borrow 1 from the tens column. Put a 1 in front of the 4 to make 14 and write a small 1 alongside the 9 in the tens column. You don't say 7 from 14, but 7 from 10 is 3, plus 4 on top gives 7, the first digit of the answer.

Ten (9 plus 1 carried) from 5 won't go so borrow again in a similar fashion. Ten from 15 is 5, or 10 from 10 is zero, plus 5 is 5.

Nine from 2 won't go, so borrow again. Nine from 10 is 1, plus 2 is 3.

Four from 7 is 3. You have your answer.

You don't have to learn or know the combinations of single-digit numbers that add to more than 10. You never subtract from any number higher than 10. This makes the calculations easier and reduces mistakes.

Try these for yourself:

a)	7,325	b)	5,417
	−4,568		− 3,179

The answers are:

a) 2,757   b) 2,238

This strategy is extremely important. If you have mastered multiplication using the simple strategies I teach in this book then you have mastered the combinations of numbers that add to ten. There are only five such combinations.

Whereas, if you had to learn the combinations of single-digit numbers that add to more than 10 there are another 20 such combinations. Using this strategy you don't need to learn any of them. To subtract 8 from 15, subtract from 10, (2) and then add the 5 for an answer of 7.

My third-grade teacher told me never to take away from a number greater than 10. There is a far greater chance of making a mistake when subtracting from numbers in the teens than when subtracting from 10. Because of practice with the tables and general multiplication using the circles, there is very little chance of making a mistake when subtracting from 10; the answers are almost automatic.

Subtraction From a Power of 10

The rule is:

Subtract the units digit from 10, then each successive digit from 9, then subtract 1 from the digit on the left.

For example,

$$\begin{array}{r} 1,000 \\ -\ 574 \\ \hline \end{array}$$

You can begin from left or right.

Let's try it from the right first. Subtract the units digit from 10.

$$10 - 4 = 6$$

This is the right-hand digit of the answer. Now take the other digits away from 9. Take 1 from the first digit.

Four from 10 is 6, 7 from 9 is 2, 5 from 9 is 4, and 1 from 1 is 0. Hence, the answer is 426.

Now let's try it from left to right: 1 from 1 is 0, 5 from 9 is 4, 7 from 9 is 2, 4 from 10 is 6. The answer is 426.

If you had to calculate 40,000 minus 2,748, this is how you would do it:

$$40000$$
$$-2748$$

Take 1 from the left-hand digit (4) to get 3, the first digit of the answer. Two from 9 is 7, 7 from 9 is 2, 4 from 9 is 5, and 8 from 10 is 2.

The answer is 37,252.

This way, we only have to subtract from numbers up to 10 then add if necessary.

The setting out is exactly the same as before. The only difference is what you say in your mind.

Try these for yourself:

a) 10,000   b) 50,000
   – 3,456   – 27,214

You would have found the answers are:

a) 6,544   b) 22,786

Subtracting Smaller Numbers

If the number we are subtracting has fewer digits than the one you are subtracting from, then add zeros before the number (at least, mentally) to make the calculation.

For instance:

$$23,000 - 46 =$$
$$23,000$$
$$- 0,046$$
$$22,954$$

You extend the zeros in front of the subtrahend (the number being subtracted) as far as the first digit that is not a zero. You subtract 1 from this digit. Three minus 1 equals 2.

Subtract each successive digit from 9 until you reach the final digit, which you subtract from 10.

The method taught in most schools has you doing exactly the same calculation, but you have to work out what you are carrying and borrowing with each step. The benefit of the method I teach here is that it becomes mechanical and can be carried out with less chance of making a mistake.

Checking Subtraction by Casting Out Nines

For subtraction, the method used to check our answers is similar to that used for addition, but with a small difference.

Let's try an example:

$$\begin{array}{r} 8,465 \\ -\ 2,897 \\ \hline 5,568 \end{array}$$

Is the answer correct?

Let's cast out the nines and see.

$$\begin{array}{rcc} 8,\!465 & 14 & 5 \\ -\ 2,\!89\cancel{7} & & 8 \\ \hline 5,568 & 24 & 6 \end{array}$$

Five minus 8 equals 6? Can that be right? Although in the actual problem we are subtracting a smaller number from a larger number, with the substitutes, the number we are subtracting is larger.

We have two options. One is to add 9 to the number we are subtracting from.

Five plus 9 equals 14. The problem then reads:

$$14 - 8 = 6$$

This answer is correct, so our calculation was correct.

Here is the option I prefer, however. Call out the problem backwards as an addition. This is probably how you were taught to check subtractions in school. Add your answer to the number you subtracted to get your original number as your check answer.

Do this with the substitutes. Add the substitutes upwards:

$$6 + 8 = 5$$
$$6 + 8 = 14 \text{ and } 1 + 4 = 5$$

Our answer is correct.

Now check these calculations for yourself and see if we have made any mistakes. Cast out the nines to find any errors. If there is a mistake, correct it and then check your answer.

a)   5,672
    − 2,596
      3,076

c)   8,542
    − 1,495
      7,147

b)   5,967
    − 3,758
      2,209

d)   3,694
    −1,236
      2,458

They were all right except for c). Did you correct c) and then check your answer by casting the nines? The correct answer is 7,047.

This method will find most mistakes for addition and subtraction problems. Use it and make it part of your calculation. It only takes a moment and it will gain you an enviable reputation for accuracy.

UNITED STATES POSTAL SERVICE

```
***** WELCOME TO *****
    EL CENTRO STATION
  LAREDO, TX  78040-5066
    03/01/05  01:05PM
```

```
Store  USPS          Trans    23
Wkstn  sys5005       Cashier  KV2QJB
Cashier's Name       ADRIANA
Stock Unit Id        ADRIANA
PO Phone Number      956-723-3643
USPS #               4879830040
```

```
 1. 37c Rept&Amph PSA          0.74
    2 @  0.37
 2. 37c Rept&Amph PSA          0.37
```

```
Subtotal                       1.11
Total                          1.11
```

```
Cash                           5.00
Change Due
  Cash                         3.89
```

Order stamps at USPS.com/shop or call
1-800-Stamp24. Go to
USPS.com/clicknship to print shipping
labels with postage. For other
information call 1-800-ASK-USPS.

Number of Items Sold: 3

THANK YOU VERY MUCH   2

            El Centro Postal Station

UNITED STATES POSTAL SERVICE

```
***** WELCOME TO *****
EL CENTRO STATION
LAREDO, TX 78040-8088
03/01/05 01:05PM

Store USPS      Trans  23
Wkstn sys5008  Cashier KV2J08
Cashier's Name    ADRIANA
Stock Unit Id     ADRIANA
PO Phone Number  956-723-3643
USPS #          4879930040

1. 37c Rep1t&mph PSA        0.74
   2 @ 0.37
2. 37c Rep1t&mph PSA        0.37

Subtotal                   1.11
Total                      1.11

Cash                       5.00
Change Due
Cash                       3.89

Order stamps at USPS.com/shop or call
1-800-Stamp24. Go to
USPS.com/clickmanship to print shipping
labels with postage. For other
information call 1-800-ASK-USPS.


Number of Items Sold: 3

THANK YOU VERY MUCH

El Centro Postal Station
```

Chapter Ten

Squaring Numbers

To square a number simply means to multiply it by itself. A good way to visualize this is, if you have a square brick section in your garden and you want to know the total number of bricks making up the square, you count the bricks on one side and multiply the number by itself to get the answer. If you have three bricks on one side of the square, you have a total of 9 bricks that make the square ($3 \times 3 = 9$). If there are 5 bricks on one side, you have 25 bricks making the square ($5 \times 5 = 25$).

Five squared means 5×5. It is written as 5^2. The small 2 written after the 5 means there are two fives multiplied together. What would a small 3 written after the 5 mean? It would mean there are three fives to be multiplied. This is a common mathematical procedure and one that everyone should know. Here are a few examples:

$5^3 = 5 \times 5 \times 5$

$4^5 = 4 \times 4 \times 4 \times 4 \times 4$

$7^3 = 7 \times 7 \times 7$

6^2 (we say, "six squared") = 36 because $6 \times 6 = 36$. We say that 36 is the square of 6.

$13^2 = 13 \times 13 = 169$

We can easily calculate this using our methods for multiplying numbers in the teens. In fact, the method of multiplication with circles is easy to apply to square numbers, because it is easiest to use when the numbers are close to each other. In fact, all of the strategies taught in this chapter make use of our general strategy for multiplication.

Squaring Numbers Ending in 5

The method for squaring numbers ending in 5 uses the same formula we have used for general multiplication. It is a fun way to play with the strategies we have already learned.

If you have to square a number ending in 5, separate the final 5 from the digit or digits that come before it. Add 1 to the number in front of the 5, then multiply these two numbers together. Write 25 at the end of the answer and the calculation is complete.

For example:

$$35^2 =$$

Separate the 5 from the digits in front. In this case there is only a 3 in front of the 5. Add 1 to the 3 to get 4:

$$3 + 1 = 4$$

Multiply these numbers together:

$$3 \times 4 = 12$$

Write 25 (5 squared) after the 12 for our answer of 1,225.

$$35^2 = 1,225$$

Let's try another:

$$75^2, \text{ or } \textbf{75 squared} =$$

Separate the 7 from the 5. Add 1 to the 7 to get 8. Eight times 7 is 56. This is the first part of our answer. Write 25 at the end of our answer and we get 5,625.

$$75^2 = 5,625$$

We can combine methods to get even more impressive answers. Let's try another:

$$135^2 =$$

Separate the 13 from the 5. Add 1 to 13 to give 14. Thirteen times 14 is 182 (use the method taught in Chapter Two). Write 25 at the end of 182 for our answer of 18,225. This can easily be calculated in your head.

$$135^2 = 18,225$$

One more example:

$$965^2 =$$

Ninety-six plus 1 is 97. Multiply 96 by 97, which gives us 9,312. Now write 25 at the end for our answer of 931,225.

$$965^2 = 931,225$$

That is impressive, isn't it? Try these for yourself:

a) $15^2 =$ e) $95^2 =$

b) $45^2 =$ f) $115^2 =$

c) $25^2 =$ g) $145^2 =$

d) $65^2 =$ h) $955^2 =$

If you used pencil and paper to calculate the answers, do them again in your head. You will find it quite easy.

The answers are:

a) 225   b) 2,025   c) 625   d) 4,225

e) 9,025   f) 13,225   g) 21,025   h) 912,025

This shortcut also applies to numbers with decimals. For instance, with 6.5×6.5 you would ignore the decimal and place it at the end of the calculation.

$$6.5^2 =$$
$$65^2 = 4,225$$

There are two digits after the decimal when the problem is written in full, so there would be two digits after the decimal in the answer. Hence, the answer is 42.25.

$$6.5^2 = 42.25$$

It would also work for $6.5 \times 65 = 422.5$.

Likewise, if you have to multiply $3\frac{1}{2} \times 3\frac{1}{2} = 12\frac{1}{4}$.

There are many applications for this shortcut.

Squaring Numbers Near 50

Our method for squaring numbers near 50 uses the same formula as for general multiplication but, again, there is an easy shortcut.

For example:

$$46^2 =$$

Forty-six squared means 46×46. Rounding upwards, $50 \times 50 = 2,500$. We take 50 and 2,500 as our reference points.

Forty-six is below 50 so we draw a circle below.

$$\text{(50)} \quad 46^2 =$$
$$-\text{(4)}$$

Forty-six is 4 less than 50, so we write a 4 in the circle. It is a minus number.

We take 4 from the number of hundreds in 2,500.

$$25 - 4 = 21$$

That is the number of hundreds in the answer. Our subtotal is 2,100. To get the rest of the answer, we square the number in the circle.

$$4^2 = 16$$
$$2,100 + 16 = 2,116 \quad \text{ANSWER}$$

Here's another example:

$$56^2 =$$

Fifty-six is more than 50 so draw the circle above.

$$+\text{(6)}$$
$$\text{(50)} \quad 56^2 =$$

We add 6 to the number of hundreds in 2,500. Twenty-five plus 6 equals 31. Our subtotal is 3,100.

$6^2 = 36$

$3,100 + 36 = 3,136$ ANSWER

Let's try one more:

$62^2 =$

(12)

(50) $62^2 =$

$25 + 12 = 37$ (our subtotal is 3,700)

$12^2 = 144$

$3,700 + 144 = 3,844$ ANSWER

Practice with these:

a) $57^2 =$

b) $51^2 =$

c) $48^2 =$

d) $39^2 =$

e) $45^2 =$

The answers are:

a) 3,249   b) 2,601   c) 2,304   d) 1,521   e) 2,025

With a little practice, you should be able to call the answer out without a pause.

Squaring Numbers Near 500

This is similar to our strategy for squaring numbers near 50.

Five hundred times 500 is 250,000. We take 500 and 250,000 as our reference points. For example:

$506^2 =$

Five hundred and six is greater than 500, so we draw the circle above. We write 6 in the circle.

$$+\;⑥$$
$$⑤⑩⑩ \quad 506^2 =$$
$$500^2 = 250,000$$

The number in the circle above is added to the thousands.

$$250 + 6 = 256 \text{ thousand}$$

Square the number in the circle:

$$6^2 = 36$$
$$256,000 + 36 = 256,036 \quad \text{ANSWER}$$

Another example is:

$$512^2 =$$
$$+\;⑫$$
$$⑤⑩⑩ \quad 512^2 =$$
$$250 + 12 = 262$$
$$\text{Sub-total} = 262,000$$
$$12^2 = 144$$
$$262,000 + 144 = 262,144 \quad \text{ANSWER}$$

To square numbers just below 500, use the following strategy.

We'll take an example:

$$488^2 =$$

Four hundred and eighty-eight is below 500 so we draw the circle below. Four hundred and eighty-eight is 12 less than 500 so we write 12 in the circle.

$$⑤⑩⑩ \quad 488^2 =$$
$$-\;⑫$$

Two hundred and fifty thousand minus 12 thousand is 238 thousand. Plus 12 squared ($12^2 = 144$).

$$238,000 + 144 = 238,144 \quad \text{ANSWER}$$

We can make it even more impressive.

For example:

$535^2 =$

(35)

(500) $535^2 =$

$250,000 + 35,000 = 285,000$

$35^2 = 1,225$

$285,000 + 1,225 = 286,225$ ANSWER

This is easily calculated in your head. We used two shortcuts—the method for squaring numbers near 500 and the strategy for squaring numbers ending in 5.

What about 635^2?

(135)

(500) $635^2 =$

$250,000 + 135,000 = 385,000$

$135^2 = 18,225$

To find 135^2 we use our shortcut for numbers ending in 5 and for multiplying numbers in the teens $(13 + 1 = 14. \ 13 \times 14 = 182)$. Put 25 on the end for $135^2 = 18,225$.

We say, "Eighteen thousand, two two five."

To add 18,000, we add 20 and subtract 2:

$385 + 20 = 405$

$405 - 2 = 403$

Add 225 to the end.

Our answer is 403,225.

Try these for yourself:

a) $506^2 =$ c) $489^2 =$

b) $534^2 =$ d) $445^2 =$

Here are the answers:

a) 256,036 b) 285,156 c) 239,121 d) 198,025

Let's solve the last problem together:

(500) $445^2 =$

   $-$(55)

250 − 55 = 195, times 1,000 equals 195,000

$55^2 = 3,025$ (use short cut for squaring numbers ending in 5)

195,000 + 3,025 = 198,025

We could have solved the entire problem by using the shortcut for squaring numbers ending in 5.

The number before the 5 is 44.

44 + 1 = 45

(50) 44 × 45 = 3,900   1,950

   $-$(6) $-$(5)   + 30

                      1,980

Affix 25 to our answer for 198,025.

So you now have a choice of methods.

Numbers Ending in 1

This shortcut works well for squaring any number ending in 1. If you multiply the numbers the traditional way you will see why this works.

For example:

$31^2 =$

Firstly, subtract 1 from the number. The number now ends in zero and should be easy to square.

$30^2 = 900$ (3 × 3 × 10 × 10)

This is our subtotal.

Secondly, add together 30 and 31—the number we squared plus the number we want to square.

30 + 31 = 61

Add this to our subtotal, 900, to get 961.

900 + 61 = 961   ANSWER

For the second step you can simply double the number we squared, 30 × 2, and then add 1.

Another example is:

$121^2 =$

121 − 1 = 120

120^2 = 14,400 (12 × 12 × 10 × 10)

120 + 121 = 241

14,400 + 241 = 14,641   ANSWER

Let's try another:

$351^2 =$

350^2 = 122,500 (use shortcut for squaring
                        numbers ending in 5)

350 + 351 = 701

122,500 + 701 = 123,201   ANSWER

One more example:

$86^2 =$

We can also use the method for squaring numbers ending in 1 for those ending in 6. For instance, let's calculate 86^2. We treat the problem as being 1 more than 85.

85^2 = 7,225

85 + 86 = 171

7,225 + 171 = 7,396   ANSWER

Practice with these:

a) $21^2 =$ e) $81^2 =$

b) $41^2 =$ f) $131^2 =$

c) $61^2 =$ g) $141^2 =$

d) $71^2 =$ h) $66^2 =$

The answers are:

a) 441 b) 1,681 c) 3,721 d) 5,041

e) 6,561 f) 17,161 g) 19,881 h) 4,356

To calculate the answers mentally, I say the first subtotal in hundreds to make the second subtotal easy to add. To square 71 in my head I say, "Seventy squared is forty-nine hundred; seventy doubled is one hundred and forty, fifty forty, plus one, fifty forty-one (5,041)."

I don't say all of this. I just say, "Forty-nine hundred, five thousand and forty . . . one."

To square 66 mentally I say, "Sixty-five squared is forty-two twenty-five"—shortcut for squaring numbers ending in 5. "Sixty-five doubled is one hundred and thirty, forty-three fifty-five, plus one, forty-three fifty-six (4,356)."

Now try all of the above problems in your head.

Numbers Ending in 9

An example is:

$$29^2 =$$

Firstly, add 1 to the number. The number now ends in zero and is easy to square.

$$30^2 = 900 \ (3 \times 3 \times 10 \times 10)$$

This is our subtotal. Now add 30 plus 29 (the number we squared plus the number we want to square):

$$30 + 29 = 59$$

Subtract 59 from 900 to get an answer of 841. (I would double 30 to get 60, subtract 60 from 900, and then add the 1.)

$$900 - 59 = 841 \quad \text{ANSWER}$$

Let's try another:

$$119^2 =$$

$119 + 1 = 120$
$120^2 = 14,400 \ (12 \times 12 \times 10 \times 10)$
$120 + 119 = 239$
$14,400 - 239 = 14,161$
$14,400 - 240 + 1 = 14,161$ ANSWER

Another example is:

$349^2 =$
$350^2 = 122,500$ (use shortcut for squaring numbers ending in 5)
$350 + 349 = 699$

(Subtract 1,000, then add 301 to get the answer.)

$122,500 - 699 = 121,801$ ANSWER

How would we calculate 84 squared?

We can also use this method for squaring numbers ending in 9 for those ending in 4. We treat the problem as being 1 less than 85.

$84^2 =$
$85^2 = 7,225$
$85 + 84 = 169$

Now subtract 169 from 7,225:

$7,225 - 169 = 7,056$ ANSWER

(Subtract 200, then add 31 to get your answer.)

Try these for yourself:

a) $69^2 =$ c) $89^2 =$

b) $79^2 =$ d) $74^2 =$

The answers are:

a) 4,761   b) 6,241   c) 7,921   d) 5,476

Let's solve b) together. To square 79 in my head I would say, "Eighty squared is sixty-four hundred. Twice eighty is one hundred and sixty.

Sixty-four hundred minus two hundred is sixty-two hundred, plus forty is sixty-two hundred and forty, plus one, sixty-two hundred and forty-one (6,241)."

Of course, you wouldn't say all of this in your head. You would just say, "Sixty-four hundred, sixty-two forty . . . one."

For c), it is probably easier to use our general multiplication formula— use 100 as a reference number and multiply 89 by 89. These strategies give you a choice of methods; it is up to you to find the easiest.

Practice these in your head until you can do them without effort.

Chapter Eleven

Short Division

If you are confident with short division you can skip this chapter. Many people, however, have difficulty calculating simple division problems. Some schools have even dropped division from their curriculum.

If you have $32 to divide among four people, you would divide 32 by 4 to find out how much each person would get. Because four eights are thirty-two, $(4 \times 8 = 32)$ then each of the four people would receive $8. This is a simple example of division. If you divided $32 among eight people, each would get $4.

If you divided 35 books among four people, they would each receive eight books and there would be three left over. We call the three left over, the remainder. We would write the calculation like this:

$$4 \overline{)35}$$
$$8\ r3$$

or like this:

$$\begin{array}{r} 8\ r3 \\ 4\overline{)35} \end{array}$$

Here is how we would divide a larger number. To divide 4,921 by 4, we would set out the problem like this:

$$4\overline{)4,921} \quad \text{or} \quad 4\overline{)4,921}$$

We begin from the left-hand side of the number we are dividing. Four (4) is the first digit on the left. We begin by asking, what do you multiply by 4 to get an answer of 4? The answer is 1, because $1 \times 4 = 4$. We would write 1 below the 4, the number we are dividing. Four divides evenly into 4, so there is nothing left over (no remainder) to carry.

We now move to the next digit, 9. What do you multiply by 4 to get 9? There is no number that gives you 9 when you multiply it by 4. We now ask, what will give an answer below 9? Two times 4 is 8, which is as close as we can get without going above. We write 2 below the 9, and the 1 left over is carried to the next digit and put in front of it.

We now divide 4 into 12. What number multiplied by 4 gives an answer of 12? The answer is 3 ($3 \times 4 = 12$). Write 3 below the 2. The last digit is less than 4 so it can't be divided. Four divides into 1 zero times with 1 remainder.

$$4 \overline{\left| 4,9^{1}21 \right.}$$
$$1,2\ 30\ r1$$

or:

$$1,2\ 30\ r1$$
$$4 \overline{\left| 4,9^{1}21 \right.}$$

The 1 remainder can be expressed as a fraction, ¼. The answer would be 1,230¼, or 1,230.25.

Using Circles

Just as our formula can be used to multiply numbers easily, it can also be used in reverse for division.

The method works best for division by 7, 8 and 9.

Here is a simple example:

$$56 \div 8 =$$
$$8 \overline{\left| 56 \right.}$$
$$②\ 7$$
$$③$$

Here is how it works. We are dividing 56 by 8. We set the problem out as above or, if you prefer, set the problem out like below. Stick to the way you have been taught.

③
  7
8 $\overline{)56}$
②

I will explain using the first layout (on page 88). We draw a circle below the 8 (the number we are dividing by—the divisor) and then ask, how many do we need to make 10? The answer is 2, so we write 2 in the circle below the 8. We add the 2 to the tens digit of the number we are dividing (5 is the tens digit of 56) and get an answer of 7. Write 7 below the 6 in 56. Draw a circle below our answer (7). Again, how many more do you need to make 10? The answer is 3, so write 3 in the circle below the 7. Now multiply the numbers in the circles.

$2 \times 3 = 6$

Subtract 6 from the units digit of 56 to get the remainder.

$6 - 6 = 0$

There is zero remainder.

The answer is 7 with 0 remainder.

Here is another example:

$65 \div 9 =$
9 $\overline{)65}$
① 7 r2
  ③

Nine is 1 lower than 10, so we write 1 in the circle below the 9. Add the 1 to the tens digit (6) to get an answer of 7. Write 7 as our answer below the 5. Draw a circle below the 7. How many more to make 10? The answer is 3. Write 3 in the circle below the 7. Multiply the numbers in the circles, 1×3 to get 3. Take 3 from the units digit (5) to get the remainder, 2. The answer is 7 r2.

Here's another example that will explain what we do when the result is too high:

$$43 \div 8 =$$

$$\begin{array}{r} 6 \\ 8\overline{)43} \\ ② \end{array}$$

Eight is 2 below 10, so we write 2 in the circle below it. Two plus 4 equals 6. We write 6 above the units digit. We now draw another circle above the 6. How many to make 10? The answer is 4 so we write 4 in the circle. To get the remainder, we multiply the two numbers in the circles and take the answer from the units digit. Our work should look like this:

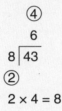

$$2 \times 4 = 8$$

We find, though, that we can't take 8 from the units digit, 3. Our answer is too high. To rectify this, we drop the answer by 1 to 5, and write a small 1 in front of the units digit, 3, making it 13.

We multiply the two circled numbers, $2 \times 5 = 10$. Take 10 from the units digit, now 13.

$$13 - 10 = 3 \text{ remainder}$$

5 r3   ANSWER

Try these problems for yourself:

a) $76 \div 9 =$ d) $62 \div 8 =$

b) $76 \div 8 =$ e) $45 \div 7 =$

c) $71 \div 8 =$ f) $57 \div 9 =$

The answers are:

   a) 8 r4   b) 9 r4   c) 8 r7

   d) 7 r6   e) 6 r3   f) 6 r3

This method is useful if you are still learning your multiplication tables and have difficulty with division, or if you are not certain and just want to check your answer. As you get to know your tables better you will find standard short division to be easy.

Chapter Twelve

Long Division by Factors

If you have $368 to divide among 16 people, you would divide 368 by 16 to find out how much each person would get.

If you don't know your 16 times tables, there is an easy way to solve the problem. Sixteen is two eights, and it is also 4 times 4. An easy way to divide by 16 is to use factors. We could divide by 4, and then divide that answer by 4. This is the same as dividing by 16, because 4 times 4 equals 16.

We could set out the problem like this:

$$\begin{array}{r|r} 4 & 368 \\ \hline 4 & 92 \\ \hline & 23 \end{array}$$

If this is upside down to the way you do short division, set it out like this:

$$\begin{array}{r|r} & 23 \\ \hline 4 & 92 \\ \hline 4 & 368 \end{array}$$

Division by numbers like 14 and 16 should be easy to do mentally. It is easy to halve a number before dividing by a factor. If you had to divide 368 by 16 mentally, you would say, "Half of thirty-six is eighteen, half of eight is four." You have a sub-total of 184. It is easy to keep track of this as you divide by 8.

Eighteen divided by 8 is 2 with 2 remainder. The 2 carries to the final digit of the number, 4, giving 24. Twenty-four divided by 8 is exactly 3. The answer is 23 with no remainder. This is easily done in your head.

A good general rule for dividing by factors is to divide by the smaller number first and then by the larger number.

The idea is that you have a smaller number to divide by the larger number.

For example, if you had to divide 3,444 by 21, you would divide by 3 first, then by 7. By the time you divide by 7 you have a smaller number to divide.

$$3,444 \div 3 = 1,148$$

$$1,148 \div 7 = 164$$

It is easier to divide 1,148 by 7 than to divide 3,444 by 7.

Division by Numbers Ending in 5

To divide by a two-digit number ending in 5, double both numbers and use factors.

For example:

$$1,085 \div 35 =$$

Double both numbers. Double 1,000 is 2,000, 2 times 85 is 170.

$$1,085 \times 2 = 2,170$$

$$35 \times 2 = 70$$

The problem is now:

$$2,170 \div 70 =$$

To divide by 70, divide by 10, then by 7 (using factors), as is shown on the next page.

$$2,170 \div 10 = 217$$
$$217 \div 7 = 31$$

This is an easy calculation. Seven divides into 21 three times (3 × 7 = 21) and divides into 7 once. Now we have the answer for our original problem:

$$1,085 \div 35 = 31 \quad \textbf{ANSWER}$$

Let's try another:

$$512 \div 35 =$$

Five hundred doubled is 1,000. Twelve doubled is 24. Hence, 512 doubled is 1,024. Thirty-five doubled is 70.

The problem is now:

$$1,024 \div 70 =$$

Divide 1,024 by 10, then by 7.

$$1,024 \div 10 = 102.4$$
$$102.4 \div 7 =$$

Seven divides into 10 once. One is the first digit of the answer. Carry the 3 remainder to the two, giving 32.

$$32 \div 7 = 4 \text{ r}4$$

We now have an answer of 14 with some remainder. We carry the 4 to the next digit, 4 to get 44.

$$44 \div 7 = 6 \text{ r}2$$

Our answer is 14.62. This is the answer to our original problem:

$$512 \div 35 = 14.62$$

You can divide numbers using factors to as many decimal places as you like.

Put as many zeros after the decimal point as you require for your answer and then add one more. This ensures that your final decimal place is accurate.

For instance, if you were dividing 736 by 21 and you need two decimal place accuracy, put three zeros after the number.

You would divide 736.000 by 21. Thus:

```
        35.047
   7 | 245.333
   3 | 736.000
```

The next section explains how to round these three decimal places to two places.

Rounding Off Decimals

To round off to two decimal places, take the third digit after the decimal. If it is below 5, you leave the second digit as it is and simply delete the third. If the third digit is 5 or more, add one to the second digit and then delete the third.

In this previous example, the third digit after the decimal is 7. Seven is higher than 5 so we round off the answer by adding 1 to the second digit, 4, to make 5.

The answer is then 35.05 to two decimal places.

The full answer to seven decimal places is 35.0476190. The decimal places begin over again, so to 13 decimal places the answer is:

35.0476190476190

To round off to 12 decimal places, the thirteenth place is a zero (below 5), so the 9 stands:

35.047619047619

To round off to 11 decimal places, the twelfth place is a nine (above 5), so the 1 is rounded off upwards to 2:

35.04761904762

To round off to 10 decimal places, the eleventh place is a two (below 5), so the 6 stands:

35.0476190476

Try these for yourself. Calculate these to two decimal places:

a) 4,356 ÷ 42 =   c) 4,173 ÷ 27 =

b) 2,355 ÷ 35 =   d) 8,317 ÷ 36 =

The answers are:

a) 103.71   b) 67.29   c) 154.56   d) 231.03

Finding a Remainder

Sometimes when we divide, we would like a remainder instead of a decimal. How do we get a remainder when we divide using factors?

The rule is:

Multiply the first divisor by the second remainder and add the first remainder.

For the above problem, we would do it like this:

```
           35 r0
       7│ 245 r1
       3│ 736
```

We begin by multiplying the corners, $3 \times 0 = 0$.

Now add the first remainder, 1. The final remainder is 1, or $\frac{1}{21}$.

One final example:

2,327 ÷ 35 =

We use 7 ÷ 5 as factors for 35.

```
           66 r3
       7│   465 r2
       5│ 2,327
```

To find the final remainder, we multiply the corners ($3 \times 5 = 15$). Now add the other remainder, 2.

15 + 2 = 17

The answer is 66 remainder 17.

Try these for yourself, calculating the remainder:

    a) 4,335 ÷ 36   b) 2,710 ÷ 24

The answers are:

    a) 120 r15   b) 112 r22

Long division by factors allows you to do many mental calculations that the average person would not attempt. I constantly calculate sports statistics mentally while a game is in progress to check a team's progress. It is a fun way to practice the strategies.

If I want to calculate the runs per at bats during a baseball game I simply divide the score by the number of at bats completed. When more than 12 at bats have been completed I will use factors for my calculations. Why not try it with your favorite sports and hobbies?

Chapter Thirteen

Standard Long Division

When we divide by prime numbers, we can't use factors to turn the problem into simple short division. (Prime numbers are numbers that have no factors, like 29.)

However, you can still use factors to solve the problem. This is done by making estimates as you go. So, for 29 we would divide by 30 (divide by 10, then by 3). For example:

24,560 ÷ 29 =

$$29 \overline{)24{,}560}$$

We can't divide 29 into 24, so we bring down the next digit into the number we are dividing. How many times do you multiply 29 to get 245? This is where people complain, how am I supposed to know the answer to that?

There is an easy way. Twenty-nine is almost 30, so we can estimate by dividing by 30. To divide by 30, we divide by 10 (easy) and then by 3 (also easy).

To divide 245 by 10, simply drop the last digit and disregard the remainder for now. The problem is now 24 divided by 3, which is easy. Three times 8 equals 24. Eight is the first digit of your answer. It goes above the line, above the 5, because we are dividing into 245.

We now multiply our answer, 8, by 29 to find our remainder. An easy way to multiply 29 by 8 is to multiply 30 by 8 and subtract 8 (30 times 8 equals 240, minus 8 is 232).

$$8 \times 29 = 232$$

Taking 232 from 245 gives a remainder of 13. Our work should look like this:

```
         8
29 | 24560
   −232
   ─────
     13
```

We now bring down the next digit in the number we are dividing (the dividend). The next digit is 6. Bring it down to the 13 remainder and we get 136. We write an x under the 6 to remind us we have used it.

We divide 136 by 29 the same way as before; divide by 10 and by 3. One hundred and thirty-six divided by 10 is 13 (dropping the remainder), and 13 divided by 3 is 4, ignoring the remainder (4 × 3 = 12). The next digit of the answer is 4. Four times 29 is 116. Take 116 from 136 to get 20.

Bring down the 0 to get 200. Divide 200 by 30 (10 × 3):

$$200 \div 10 = 20$$

$$20 \div 3 = 6$$

This is the last digit of our answer.

$$6 \times 29 = 174$$

Subtract 174 from 200 to get the remainder, 26. The completed problem looks like this:

```
        846
29 | 24560
   −232ˣˣ
   ─────
     136
    −116
    ─────
     200
    −174
    ─────
      26    remainder
```

846 r26   ANSWER

The general rule for standard long division is this:

Round off the divisor to the nearest ten, hundred or thousand to make an easy estimate.

⇨ If you are dividing by 31, round off to 30 and divide by 3 and 10.

⇨ If you are dividing by 87, round off to 90 and divide by 9 and 10.

⇨ If you are dividing by 321, round off to 300 and divide by 3 and 100.

⇨ If you are dividing by 487, round off to 500 and divide by 5 and 100.

⇨ If you are dividing by 6,142, round off to 6,000 and divide by 6 and 1,000.

This way, you are able to make an easy estimate and then proceed in the usual way to correct as you go.

Let's do another example:

13,570 ÷ 317 =

You set out the problem in the usual way:

317 | 13,570

We will round off 317 to 300 for our estimates and use factors of 100 and 3.

You can't divide 300 into 1 or 13. You can't divide 300 into 135 but it will divide into 1,357. How many times?

To divide by 300, we divide 1,357 by 100 and then by 3.

To divide 1,357 by 100 you move the decimal two places to the left or, easier still for our purposes, drop the last two digits. The problem is now 13 ÷ 3.

The answer, of course, is 4 remainder 1. We are not concerned about the remainder at this stage, just the answer, 4. We write 4 as the first digit of the answer.

```
            4
317 |13,570
```

We divided 1,357 by 317 to get our answer so we write 4 above the final digit of the number we divided, 7.

Now multiply 317 by 4 to get the actual answer and remainder.

$$317 \times 4 = 1,268$$

Write 1,268 below 1,357 and subtract.

$$1,357 - 1,268 = 89$$

Our calculation thus far should look like this:

```
            4
317 | 13570
      1268
       89
```

Now bring down the next digit of the number we are dividing. The next digit is 0. Our new working number is 890. We now have to divide 890 by 317. Divide by 100 equals 8. Divide by 3 equals 2. Place this digit above the zero in the problem.

We multiply 317 by 2 to find the remainder.

$$2 \times 317 = 634$$
$$890 - 634 = 256$$

Two hundred and fifty-six is our remainder.

Our final calculation looks like this:

```
             42
317 | 13570
      1268ˣ
       890
       634
       256    remainder
 42 r256    ANSWER
```

If we want our answer to be expressed as a decimal we continue the operation. The general rule when dividing is to write one more zero after the decimal than the number of decimal places you want in the answer.

If we would like our answer to one decimal place, divide 13,570.00 by 317 and round off:

$$
\begin{array}{r}
42.8 \\
317\overline{\smash{)}13{,}570.00} \\
12{,}68^{\text{x x}} \\
\hline
890 \\
634 \\
\hline
2560 \\
2536 \\
\hline
24 \quad \text{remainder}
\end{array}
$$

In this case, even if we continue to another decimal place we can see that if we bring down another zero to make the next step 240 ÷ 317 we would have an answer of less than zero. That would give us 42.80, so our answer of 42.8 is certainly accurate to one decimal place.

To divide by 317, the largest number we used to divide by was 3. That made the calculation simple.

Thus, we can say that long division is not difficult.

Which numbers would you use as factors to divide by these numbers?

a) 78   c) 723

b) 289   d) 401

Here are the numbers I would use for my estimates:

a) 8 × 10 (80)   c) 7 × 100 (700)

b) 3 × 100 (300)   d) 4 × 10 (400)

What estimate would you use to divide by 347? Three hundred and forty-seven is closer to 300 than to 400, but neither is really satisfactory. An easier option would be to double both the divisor and the dividend.

For example, to divide 33,480 by 347, double both numbers for the same answer. Doubling 33,480 we get 66,960, 347 doubled is 694.

The problem is now 66,960 divided by 694. We use 700 as our estimate. You could easily make a close mental estimate by dividing 67,000 by 700.

Dividing by 100 we get 670, then dividing 670 by 7 we get 9 with 4 remainder. Divide 7 into 40 to get almost 6. Our estimate is 96.

Performing the actual calculation we get 96.48.

At one time, I was taking part in a U.S. government program to teach students effective methods of learning and I made the statement that I taught students how to use factors to divide by prime numbers. That was too much for one teacher, who challenged me to demonstrate.

I showed the teachers of the particular school the methods I teach in this chapter. The teacher who made the challenge said: "You know, I have always done long division that way, but it never occurred to me to explain it that way to my students."

Chapter Fourteen

Direct Division

If you are comfortable with short division, long division should be easy. If you are dividing by a number that is not a prime—which means the number can be broken down into factors—the problem is easy. Standard long division is also easy if you use the principle of factors to estimate. Here is another alternative which can be used for division by two-digit and three-digit numbers.

Division by Two-Digit Numbers

Let's take an example:

$$2,590 \div 73 =$$

Firstly, we round off 73 to 70 and divide the number by 10 and 7, making corrections as we go for the 3.

Divide the number by 10. This will place a decimal point in the answer.

$$2,590 \div 10 = 259.0$$

We now divide 259.0 by 7, correcting for the 3 as we go.

$$7^3 \overline{\left) 25\,^49.\,^5000 \right.}$$
$$\phantom{7^3 \overline{\left)}} 3\ 5$$

Seven divides into 25 three times ($3 \times 7 = 21$) with 4 remainder. Three is the first digit of the answer. We carry the remainder as in standard division. Carrying the 4 to the next digit, 9, gives a working total of 49. We now correct this by multiplying the previous digit of the answer (3) by the units digit (3) of the actual divisor, 73. That is, 3 times 3 equals 9. Subtract 9 from our working total (49) to give an answer of 40. We now divide 7 into 40. The answer is 5, as $5 \times 7 = 35$. Five is the second digit of the answer. We carry 5 remainder to the next digit and 50 becomes our next working number.

$$7^{3}\big|\ 25\ ^{4}9.\ ^{5}000$$
$$35$$

Multiply our last answer, 5, by the units digit of the divisor, 3, to get 15. Subtract 15 from 50 to get 35. We divide 35 by 7 to get an exact answer of 5. We have no remainder, and that gives us a problem because we have to subtract our answer multiplied by 3 from our next working number. So we drop the answer by 1 to 4 with 7 remainder.

$$7^{3}\big|\ 25\ ^{4}9.\ ^{5}0\ ^{7}00$$
$$35.\ 4$$

Multiply this answer, 4, by the units digit, 3, to get 12. Subtract 12 from 70 to get 58. Fifty-eight divided by 7 gives 8 with 2 remainder. The 2 carried will give us a new working number of 20. Is this big enough? We will have to subtract $3 \times 8 = 24$ from 20. Our answer is too high again so we decrease it by 1 to get 7. Fifty-eight divided by 7 gives us 7 with 9 remainder. Write 7 and carry the 9. We have a working number of 90. Next, $7 \times 3 = 21$ and $90 - 21 = 69$. This is acceptable.

Dividing 69 by 7 we get 9 with 6 remainder. 9 is the next digit of the answer.

$$7^{3}\big|\ 25\ ^{4}9.\ ^{5}0\ ^{7}0\ ^{9}0\ ^{6}0$$
$$35.479$$

With practice, all this can be calculated mentally.

Let's try another.

$$2{,}567 \div 31 =$$

Thirty is 3 times 10, so we divide by 10, then by 3, correcting for the units digit.

$$2{,}567 \div 10 = 256.7$$

Three divides into 25 eight times with 1 remainder. Eight is the first digit of our answer. The 1 is carried to the next digit, to give 16.

$$3^{1} \overline{\smash{)}2\,5^{1}6\,.\,7}$$
$$8$$

Correcting for the units digit, we multiply our last answer, 8, by the units digit, 1. Eight times 1 is 8. Subtract 8 from the new working number, 16, for an answer of 8.

We now divide 8 by 3. The answer is 2 with 2 remainder. Carry the 2 remainder to the next digit. We have a new working number of 27. We still need to correct this with respect to the units digit.

$$3^{1} \overline{\smash{)}2\,5^{1}6\,.^{2}7}$$
$$8\,2$$

The previous digit of the answer was 2. Multiply this by the units digit of the divisor. Two times 1 is 2, 27 minus 2 equals 25. Divide 25 by 3 to get 8 with 1 remainder.

$$3^{1} \overline{\smash{)}2\,5^{1}6\,.^{2}7^{1}0\,0}$$
$$8\,2.8$$

Multiply this digit of the answer, 8, by the units digit, 1, to get an answer of 8. Subtract 8 from the new working number, 10. Ten minus 8 equals 2. Three divides into 2 zero times. The next digit of the answer is zero.

This gives us an answer correct to one decimal place, 82.8.

Try the following problems for yourself. If you like, you can write out all your work. Try to mentally calculate some answers then write the answer only.

a) $368 \div 71 =$ d) $549 \div 61 =$

b) $236 \div 43 =$ e) $1,234 \div 41 =$

c) $724 \div 61 =$

The answers are:

a) 5.18   b) 5.488   c) 11.869   d) 9   e) 30.09756

A Reverse Technique

If you have a high units digit, you can solve the problem using a reverse procedure.

For example:

$2,590 \div 69 =$

For 69, substitute $(70 - 1)$.

We divide by 10 and then by 7, making corrections as we go.

$$7^{-1} \overline{\left| 25^{4}9.000 \right.}$$
$$3$$

Seven divides into 25 three times ($3 \times 7 = 21$) with 4 remainder. We carry the 4 as before to give a working number of 49. We now multiply our answer, 3, by the 1 which we regard as our units digit. The answer is 3. We *add* this answer to our working number to get 52. Next, divide 52 by 7 to get 7 with 3 remainder. Write the 7 and carry the 3. We have a working number of 30.

$$7^{-1} \overline{\left| 25^{4}9.^{3}0 \right.}$$
$$3\,7$$

Now multiply our last answer, 7, by 1, which equals 7. Add 7 to 30 to get 37. Thirty-seven divided by 7 is 5 remainder 2. Write 5 and carry 2. We have a working number of 20. Add 5 × 1 = 5 and we get 25. Twenty-five divided by 7 equals 3 remainder 4. Carry the 4 to get 40. Add 3 to get 43. Forty-three divided by 7 is 6, the next digit of the answer. You can take the answer to as many decimal places as you wish. The completed problem up to three decimal places looks like this:

$$7^{-1}\overline{|25\overset{4}{9}.\overset{3}{0}\overset{2}{0}\overset{4}{0}}$$
$$3\ 7.5\ 3\ 6$$

Try these examples for yourself:

a) 2,671 ÷ 41 =   c) 3,825 ÷ 62 =

b) 3,825 ÷ 58 =   d) 2,536 ÷ 39 =

The answers are:

a) 65.146   b) 65.948   c) 61.69   d) 65.0256

If you round up the original divisor to get the substitute divisor, you then *add* the corrections to your working numbers. If you round down the original divisor to get the substitute divisor, you *subtract* the corrections from your working numbers.

An easy way to remember if you add or subtract the correction is to think of dividing 15 gifts among 9 people or among 11 people. Which division would give the greater remainder? If you divided by 10, in the first instance, you would have to add 1 to correct for dividing by 9. In the second instance, you would subtract 1 to correct for dividing by 11.

Here is an example of a complication using this method and how to deal with it.

Let's try 2,536 ÷ 39. Here is how I set out the problem:

40
+1
39 | 2,536

I write the divisor, 39, then write "+1" above to make our working divisor of 40. (The plus 1 tells me to add plus 1 times the last digit of the answer for the correction.)

To divide by 40, we divide by 10, then by 4.

Two thousand, five hundred and thirty-six divided by 10 is 253.6. Now divide by 4, making our corrections as we go.

        40
        +1
        39 | 253.6000

Four divides into 25 six times with a remainder of 1. Carry the 1 to the next digit, 3, to make 13.

        40
        +1
                 1
        39 | 25 3.6000
             6

Now make the correction. Six times +1 is +6. Add 6 to our working number of 13 to get 19. Nineteen divided by 4 is 4 with 3 remainder. Write the 4 and carry the 3 to get 36 as our next working number.

        40
        +1
                 1 3
        39 | 25 3. 6000
             6 4.

Four times +4 is +4. Thirty-six plus 4 equals 40. Forty divided by 4 is 10.

Now we have a problem. Ten is not a valid digit so we know our last answer was too low. Raise it from 4 to 5.

        40
        +1
                 1 −1
        39 | 25 3. 6000
             6 5.

Four divides into 19 five times, with −1 remainder.

(In other words, 5 times 4 is 20. Nineteen, the working number we are dividing, is 20 − 1.)

109

When we carry 1 to the next digit, it represents a 10. (Two represents 20, 3 represents 30 and so on. We are multiplying the carried number by 10.)

Multiply the last digit of our answer, 5, by +1 to get +5. The next working number is 6 (minus 10 carried) plus 5 is 11, minus the 10 carried is 1.

Four divides into 1 zero times with 1 remainder.

$$\begin{array}{r} 40 \\ +1 \\ 39\,\big|\,25\,\overset{1}{3}.\ \overset{-1}{6}\,\overset{1}{0}00 \\ \hline 6\,5.\ 0 \end{array}$$

The next working number is 10, plus zero times +1 = 10.

Four divided into 10 goes twice with 2 remainder.

$$\begin{array}{r} 40 \\ +1 \\ 39\,\big|\,25\,\overset{1}{3}.\ \overset{-1}{6}\,\overset{1}{0}\,\overset{2}{0}0 \\ \hline 6\,5.\ 0\,2 \end{array}$$

We now have a working number of 20 plus 2 times 1, giving us 22.

Four divided into 22 goes 5 times with 2 remainder.

$$\begin{array}{r} 40 \\ +1 \\ 39\,\big|\,25\,\overset{1}{3}.\ \overset{-1}{6}\,\overset{1}{0}\,\overset{2}{0}\,\overset{2}{0} \\ \hline 6\,5.\ 0\,2\,5 \end{array}$$

Now our working number is 20, plus 5 gives us 25.

Four divided into 25 gives us 6 with 1 remainder.

$$\begin{array}{r} 40 \\ +1 \\ 39\,\big|\,25\,\overset{1}{3}.\ \overset{-1}{6}\,\overset{1}{0}\,\overset{2}{0}\,\overset{2}{0}\,\overset{1}{0} \\ \hline 6\,5.\ 0\,2\,5\,6 \end{array}$$

Ten plus 6 is 16, divided by 4 gives us 4. We can see our answer, 65.0256, is accurate to 4 decimal places.

When we get an answer of 10 during our division, we know we have to raise the last digit of our answer by 1. We then get a negative remainder to carry, which represents 10 times the carried number. Ignore it until you have made the correction by multiplying the previous digit of your answer by the correction factor. It is easier to subtract a multiple of 10 at the end than to subtract at the beginning and work with negative numbers.

Division by Three-Digit Numbers

Division by three-digit numbers is similar to division by two-digit numbers. For example:

$$45,678 \div 321 =$$

We set out the problem the same way as we did for the two-digit division.

$$321 \mid 45,678$$

Firstly, we divide by 300. This means we will divide by 100, then by 3.

To divide by 100, we move the decimal point two places to the left. That gives us 456.78.

Now we divide by 3, making corrections as we go.

Three divides into 4 one time with 1 remainder, so 1 is the first digit of the answer. Write 1 below the 4. Carry the 1 remainder to the next digit to make a working number of 15.

$$321 \mid 4\overset{1}{}56.78$$
$$1$$

We now multiply our answer, 1, by the second digit of the divisor, 2. One times 2 is 2. Subtract 2 from our working total, 15, for an answer of 13. We now divide 13 by 3. Thirteen divided by 3 is 4 with 1 remainder. Four is the next digit of the answer. Carry the 1 remainder to get a new working number of 16.

$$321 \mid 4\overset{1}{}5\overset{1}{}6.78$$
$$1\,4$$

111

We now multiply our answer, 4, by the second digit of the divisor:

$$4 \times 2 = 8$$

We also multiply our previous answer, 1, by the third digit of the divisor, 1. Add these two answers ($8 + 1 = 9$) and subtract from our working number. Our working number is 16. So, $16 - 9 = 7$.

We divide 7 by 3 to get the next digit of the answer. Seven divided by 3 equals 2 with 1 remainder.

We carry the remainder to the next digit, 7, to give a new working number of 17.

$$321 \mid 4 \overset{1}{5} \overset{1}{6}. \overset{1}{7} 8$$
$$1\ 4\ 2$$

We multiply crossways the last two digits of the answer with the last two digits of the divisor. Then add these two numbers together.

$$2 \times 2 = 4$$
$$4 \times 1 = 4$$
$$4 + 4 = 8$$

Subtract 8 from our working number, 17 ($17 - 8 = 9$). Divide 9 by 3.

We know we need a remainder to subtract from. If we carry nothing, will the next digit, 8, be enough? Yes, our next cross multiplication will add up to 8 ($8 \div 3 = 2$ r2, then $3 \times 2 = 6$, $2 \times 2 = 2$, $6 + 2 = 8$), but there would be no remainder left for the final step. So we drop our answer by 1 to give 3 remainder. Let's continue:

$$9 \div 3 = 2\ r3$$
$$321 \mid 4 \overset{1}{5} \overset{1}{6}. \overset{1}{7} \overset{3}{8}$$
$$1\ 4\ 2.\ 2$$

We now have a working number of 38. Multiply crossways, then add.

$$2 \times 2 = 4$$
$$2 \times 1 = 2$$
$$4 + 2 = 6$$
$$38 - 6 = 32$$

We divide 32 by 3 and get an answer of 9 with 5 remainder. (We can't get 10 as an answer—the answer must be a single-digit number.) We now know we will have a larger number to subtract when we multiply crossways because of our 9 answer.

$$321 \, | 4 \, \overset{1}{5} \, \overset{1}{6}. \, \overset{1}{7} \, \overset{3}{8} \, \overset{5}{0}$$
$$1 \, 4 \, 2. \, 2 \, 9$$

Our next working number is 50. Multiply crossways:

$9 \times 2 = 18$

$2 \times 1 = 2$

$18 + 2 = 20$

$50 - 20 = 30$

We know 30 is all right because it still allows a cross multiplication to be subtracted.

$30 \div 3 = 9 \, r3$

$$321 \, | 4 \, \overset{1}{5} \, \overset{1}{6}. \, \overset{1}{7} \, \overset{3}{8} \, \overset{5}{0} \, \overset{3}{0}$$
$$1 \, 4 \, 2. \, 2 \, 9 \, 9$$

We now have a working number of 30. Multiply crossways:

$9 \times 2 = 18$

$9 \times 1 = 9$

$18 + 9 = 27$

$30 - 27 = 3$

Three won't divide by 3 and give a remainder so the next digit is zero.

We will leave the problem there for an answer of 142.2990, which is accurate to four decimal places. You can continue for as many decimal places as you wish.

Isn't this easier than conventional long division?

Here are some problems to try for yourself:

a) $7{,}120 \div 312 =$ b) $4{,}235 \div 213 =$

The answers are on the next page.

a) 22.82   b) 19.88

Try your own examples and check your answers with a calculator. In Chapter Sixteen we will learn a quick and easy method for checking problems like these.

Chapter Fifteen

Division by Addition

This method of division is great for dividing by numbers a little lower than powers of 10 or multiples of powers of 10. It also helps to illustrate what division is all about.

How many times will 9 divide into 10? One time with 1 remainder. That is, for every 10, 9 will divide once with 1 remainder. So, for 20, 9 will divide twice with 2 remainder. For 40, 9 will divide four times with 4 remainder. For every 10, 9 divides once with 1 remainder.

If you have a handful of dollars you can buy one item costing 90¢ for each dollar and have 10¢ change. If you have enough dollars, you can buy some extra items out of the change. This brings us to a new and easy way of calculating some long division problems. If you are dividing by 90, divide by 100 (a dollar) and give back the change.

For instance, if you are buying a drink for 95¢ and you have $1.20, you would hand over a dollar and keep the 20¢ in your pocket, plus you would get 5¢ change, making 25¢ left over, or 25¢ remainder.

So, you could say, 95 divides into 120 one time with 25 remainder.

Let's try an example:

$$\begin{array}{r} 100 \\ 96\overline{)234} \\ \textcircled{4} \end{array}$$

We are dividing 234 by 96. We write the problem in the conventional way but we draw a circle below the divisor, 96, and write 4 in the circle. (How many to make 100?)

Now, instead of dividing by 96, we divide by 100. How many times will 100 divide into 234? Two times with 34 remainder. Write 2 as the answer.

We have our 34 remainder plus another 4 remainder for every 100. We have two hundreds so our remainder is 2 times 4, which is 8, plus the 34 from 234, giving us a total of 42 remainder.

We would set out the problem like this:

```
100     2
    ____
 96 | 234
 ④     8+
    ____
       42     remainder
```

We don't add the hundreds digit because we have finished with it. In effect, what we are saying is, 100 divides into 234 twice, with 34 left over. Because we are really dividing by 96, we also have 4 remainder for each hundred it divides into.

If we were purchasing articles for 96¢ and we had $2.34 in our pocket, we would hand $2.00 to the seller and keep the 34¢ in our pocket. We would get 8¢ change which, added to the 34¢, makes a total of 42¢.

Let's try another.

$$705 \div 89 =$$

We set it out:

```
100
    ____
 89 | 705
 ⑪
```

Eighty-nine is 11 less than 100 so we write 11 below the 89.

How many hundreds are there in 705? Obviously 7, so we write 7 as the answer.

What is our remainder? For every hundred we have 11 remainder. We have 7 hundreds so we have 7 times 11 remainder. Seven times 11 is 77, plus the 5 left over from 705 gives us 82 (77 + 5 = 82).

```
100    7
 89 | 705
(11)   77
       82
```

Again, we add the 77 to the 5, not 705 as we have finished with the hundreds and are only concerned with the 5 remainder.

Let's take the answer to two decimal places.

```
100    7.
 89 | 705.000
(11)   77 ˣ
       820
```

We now divide 89 into 820.

We have 8 hundreds so 8 is the next digit of the answer. Eight times 11 equals 88.

```
100    7.8
 89 | 705.000
(11)   77 ˣ
       820
        88
```

Eighty-eight plus 20 equals 108 remainder. This is obviously too high as it is greater than our divisor so we add 1 to our answer. Our calculation now looks like this:

```
100    7.9
 89 | 705.000
(11)   77 ˣ
       820
        99
       119
```

Because we had to increase the answer by one, we subtract one times the working divisor, 100, from our remainder. We cross out the hundreds digit of the remainder to get an actual remainder of 19.

We bring down the next zero to make it 190.

We can see that 89 will divide twice into 190 so we can simply write 2 as the next digit of the answer. (The 90 part of the number is already greater than our divisor.)

(If we couldn't see at a glance that the answer is 2, we could write 1 as the next digit because there is 1 hundred in 190. One times 11 gives us 11 to add to the 90 of 190 to make 101 remainder. Seeing our divisor is only 89, we can't have a remainder of 101.)

$$
\begin{array}{r}
100 \quad 7.92 \\
89\overline{\smash{)}705.000} \\
\textcircled{11} \quad 77^{\,XX} \\
\hline
820 \\
99 \\
\hline
1{,}190 \\
22 \\
\hline
112 \quad \text{remainder}
\end{array}
$$

Because we increased the digit of the answer by 1, we subtract 1 times 100 from the remainder. One hundred and twelve minus 100 is 12. Bring down the final zero to make 120.

Eighty-nine divides once into 120 making the answer 7.921. Seeing we only need our answer to two decimal places, we can round off to 7.92. You could easily solve this entirely in your head.

Try calculating all of these in your head, just giving the answer with remainder.

a) $645 \div 98 =$ c) $234 \div 88 =$

b) $2{,}345 \div 95 =$ d) $1{,}234 \div 89 =$

The answers are:

a) 6 r57   b) 24 r65   c) 2 r58   d) 13 r77

Easy, weren't they?

This method works well when dividing by numbers just below a power of 10, or a multiple of a power of 10, but it can be made to work for other numbers as well.

Dividing by Three-Digit Numbers

Example:

$$23,456 \div 187 =$$

We set out the problem like this:

```
 200
187 | 23,456
⑬
```

We use a working divisor of 200 because 187 is 200 minus 13.

We make our first calculation. Two hundred divides into 234 one time, so the first digit of the answer is 1. We write the 1 over the 5.

We multiply the answer, 1, by the number in the circle, 13, to get an answer of 13. Write 13 under 234 and add it to the 34.

$$34 + 13 = 47$$

```
 200      1
187 | 23,456
⑬       13
        ___
        47
```

Now bring down the next digit, 5, to make 475.

Divide 475 by 200. Two hundred divides into 400 twice, so 2 is the next digit of the answer.

```
 200     12
187 | 23,456
⑬       13ˣ
        ____
        475
```

Multiply 2 by 13 to get 26.

$$75 + 26 = 101$$

Bring down the next digit, 6.

```
200      12
187 | 23,456
(13)    13ˣˣ
        ———
        475

         26
        ———
       1,016
```

Divide 1,016 by 200. Two hundred divides into 1,000 five times, so the next digit of the answer is 5. Five times 13 is 65. Sixty-five plus 16 equals 81; this is our remainder.

```
200      125
187 | 23,456
(13)    13ˣˣ
        ———
        475

         26
        ———
       1,016

         65
        ———
         81    remainder
```

The answer is 125 with 81 remainder.

It is much easier to multiply the 13 by each digit of the answer than to multiply 187 by each digit.

The following example will show what to do when you have a remainder when you divide by your working divisor.

$$4,567 \div 293 =$$

We set out the problem like this:

```
300
293 | 4,567
(7)
```

We divide 400 by 300 to get an answer of 1 with 100 remainder. I write a small 1 to indicate the remainder.

One times 7 is 7, so we add 7 to 56, plus the 100 carried to get our remainder.

```
300     1
293 | 4,567
 ⑦    ¹07
      ─────
      163
```

Bring down the next digit of the number we are dividing. We now divide 1,637 by 300.

Three hundred divides into 1,600 five times with 100 remainder. We follow the same procedure.

Thirty-seven plus 135 gives us our remainder of 172. We had to carry a remainder twice in this calculation.

```
300      15
293 | 4,567
 ⑦    ¹07ˣ
      ─────
      1,637
       ¹35
      ─────
       172    remainder
```

Here is another example:

45,678 ÷ 378 =

Three hundred and seventy-eight is 22 less than 400 so we can easily use this method.

Here is how we set out the problem:

```
400
378 | 45,678
 ㉒
```

We use 400 as our working divisor.

The first calculation is easy. Four hundred divides once into 456.

```
400      1
378 | 45,678
(22)    22
        ──
        78
```

One times 22 is 22. Add 22 to 56 to get 78.

Bring down the next digit of the number we are dividing, 7.

```
400      1
378 | 45,678
(22)    22ˣ
        ───
        787
```

Four hundred divides once into 787 but 787 is almost 800 and we are not actually dividing by 400, but by 378. We can see the answer is probably 2. Let's try it.

```
400      12
378 | 45,678
(22)    22ˣ
        ───
        787

         44
        ───
        131
```

Two times 22 is 44. Eighty-seven plus 44 is 131. Subtract 100 to get 31 because we have added 1 to our answer to make 2 instead of 1. Does this make sense? Yes, because we are finding the remainder from subtracting twice 378 from 787. The answer must be less than 100. Now bring down the next digit, 8.

```
400      12
378 | 45,678
(22)    22ˣˣ
        ───
        787

         44
        ───
        318
```

Four hundred will not divide into 318. We now check with our actual divisor. Three hundred and seventy-eight will not divide into 318 so the next digit of the answer is zero and the 318 becomes the remainder.

```
400      120
378 | 45,678
 22      22^XX
         787

          44
         318    remainder
```

How would you solve this problem?

1,410 ÷ 95 =

You could solve this two ways using this method. Let's see. Here is the first way.

```
100
 95 | 1,410
  5
```

You could ask yourself, "How many times does one hundred divide into fourteen hundred?" Fourteen times. Write down 14 as the answer.

Now the remainder. Fourteen times 5 is 70. (Fourteen is 2 times 7. Multiply 5 by 2—10—and by 7, 70.) Seventy plus the remainder from 1,410 (10) makes 80. The answer is 14 with 80 remainder.

```
100      14
 95 | 1,410
  5      70
         80    remainder
```

Here is the second way:

```
100      14
 95 | 1,410
  5       5^X
         460

          20
          80    remainder
```

One hundred divides into 141 one time with 41 remainder.

One times 5 is 5, plus 41 remainder makes 46. Bring down the next digit, 0, to make 460.

One hundred divides into 460 four times with 60 remainder. Four times 5 is 20, plus the 60 remainder makes a remainder of 80.

The first method is easier, don't you think, and more suited to mental calculation? Try calculating the problem both ways in your head and see if you agree.

Possible Complications

Here is an interesting example which demonstrates possible complications using this method.

$$3,456 \div 187 =$$

We set out the problem:

```
    200
187 | 3,456
(13)
```

We divide 200 into 300 of 345. The answer is 1, with 100 remainder. We write it like this:

```
    200      1
187 | 3¹,456
(13)
```

We write a small 1 above the hundreds digit (3 is the hundreds digit of 345) to signify 100 remainder.

Now multiply 1 times 13 equals 13. Add this to the 45 of 345, plus our 100 carried.

```
    200      1
187 | 3¹,456
(13)     13
       -----
        158
```

What we are saying is this: if we have $345 in our pocket and purchase something for $187, we can hand over $200 to the seller and keep the other $145 in our pocket. We will be given $13 change which, added to the money in our pocket, makes $158 we have left.

We bring down the 6 for the final part of the calculation. Now we have 1,586 divided by 200.

Two hundred divides into 1,500 seven times with 100 left over. Don't forget to write this 100 remainder as a small 1 in the hundreds column. Because we have 7 times 13 remainder, plus the 186 from 1,586, we can see we can raise our answer by 1 to 8. Eight times 13 is easy. Eight times 10 is 80, plus 8 times 3 is 24, 104.

$$
\begin{array}{ll}
200 & 18 \\
187\,\big|\,3^{1}\!,456 & \\
\textcircled{13} \quad 13^{X} & \\
\overline{1,5^{1}86} & -200 \\
104 &
\end{array}
$$

Because the extra 1 we have added to the answer accounts for another 200 we are dividing by, this must be subtracted from the remainder. I write "–200" to the side of my work to remind me.

$$
\begin{array}{ll}
200 & 18 \\
187\,\big|\,3^{1}\!,456 & \\
\textcircled{13} \quad 13^{X} & \\
\overline{1,5^{1}86} & -200 \\
104 & \\
\overline{\,90} &
\end{array}
$$

We add 186 to 104 to get 290. Now subtract the 200 written at the side to get our final remainder of 90. That is as complicated as it can get but you will find it becomes easy with practice. As long as you can keep track of what you are doing it won't be difficult. Practice some problems for yourself and you will find it will become very easy.

Can we use this method to divide 34,567 by 937? Although 937 is not far from 1,000, we still have a larger difference—one that is not so easy to multiply.

Let's try it.

```
1,000
937 |34,567
63
```

The first calculation would be, 3,000 divided by 1,000. The answer is obviously 3. This is the first digit of the answer.

Now we have to multiply our circled number, 63, by 3.

Three times 60 is 180, and 3 times 3 is 9; the answer is 189. Write 189 below 3456 and add it to 456, for the remainder.

$$456 + 189 = 645$$

```
1000      3
937 |34567
63      189
        645
```

Now bring down the next digit, 7.

```
1000      3
937 |34567
63      189ˣ
        6457
```

We now have to divide 6,457 by 1,000.

Six thousand divided by 1,000 is 6. Now we multiply 63 by 6. Is this difficult? No. Six times 60 is 360, plus $3 \times 6 = 18$, giving us 378.

Add this to 457 to get our remainder of 835.

```
1000      36
937 |34567
63      189ˣ
        6457
        378
        835   remainder
```

So, 34,567 divided by 937 is 36 with 835 remainder.

Let us continue the calculation to two decimal places.

```
 1000      36
  937 | 34567.000
 (63)    189ˣ
         ____
         6457

          378
         ____
          835
```

We add one more zero (or decimal place) than we need for the answer.

Bring down the first zero to make 8,350. How many times will 1,000 divide into 8,000? Eight times, so 8 is the next digit of the answer.

```
 1000      36.8
  937 | 34567.000
 (63)    189ˣ ˣ
         ____
         6457

          378
         ____
          8350
```

Eight times 63 is 504. (Eight times 60 is 480 and 8 times 3 is 24. Four hundred and eighty plus 24 equals 504.)

$$350 + 504 = 854$$

```
 1000      36.8
  937 | 34567.000
 (63)    189ˣ ˣ
         ____
         6457

          378
         ____
          8350

          504
         ____
          854
```

Bring down the next digit, 0, to make 8,540, and divide 8,540 by 1,000 to get 8 again.

We know that 8 multiplied by 63 is 504, so we add this to 540 to get 1,044.

```
1000        36.88
 937 | 34567.000
 63    189ˣ ˣˣ
       ────
       6457
        378
       ────
       8350
        504
       ────
       8540
        504
       ────
       1044
```

This is clearly wrong because we have a remainder higher than our divisor so we have to increase our last digit of the answer by 1. We cross out the last digit, 8, and raise it to 9. Nine times 63 is 567. (Nine times 60 equals 540, 9 times 3 is 27, and 540 plus 27 equals 567.)

$$540 + 567 = 1,107$$

```
1000        36.89
 937 | 34567.000
 63    189ˣ ˣˣ
       ────
       6457
        378
       ────
       8350
        504
       ────
       8540
        567
       ────
       1107
```

We subtract 1,000 because we raised the last digit of our answer by 1. Bring down the final zero to calculate the third decimal place. One thousand and seventy divided by 1,000 is 1. This gives us an answer of 36.891. We want an answer to two decimal places: 36.891 rounds off to 36.89. Our calculation is complete.

Again, it is much easier to multiply 63 by each digit of our answer than to multiply 937. The full calculation looks like this:

```
1000        36.891
 937 | 34,567.000
 (63)    189ˣ ˣˣ
         6457
          378
         8350
          504
         8540
          567
         1070
```

Try these problems for yourself. Calculate for a remainder, then to one decimal place.

a) $456 \div 194 =$ c) $5,678 \div 186 =$

b) $6,789 \div 288 =$ d) $73,251 \div 978 =$

How did you do? Here are the problems fully worked.

a)

```
200     2
194 | 456
(6)    12
       68    remainder
```

```
200     2.35
194 | 456.00
(6)    12 ˣˣ
       680
        18
       980
        30
        10    remainder
```

The answer is 2.4 to one decimal place.

b)

```
      300      23
      288 | 6,789
     (12)    24ˣ
            ─────
            1,029

              36
            ─────
             165     remainder
```

```
      300       23.57
      288 | 6,789.00
     (12)    24ˣ ˣˣ
            ───────
            1,0¹29

              36
            ───────
            1,6¹50

               60
            ───────
            2,100

               84
             ─────
               84
```

The answer is 23.6 to one decimal place.

c)

```
      200      30
      186 | 5,678
     (14)    42ˣ
            ─────
             098     remainder
```

```
      200      30.52
      186 | 5678.00
     (14)    42ˣ ˣˣ
            ───────
            0980

              70
            ───────
            5¹00

              28
            ─────
             128
```

The answer is 30.5 to one decimal place.

To multiply by 14 we simply multiply by 7 and double the answer (2 × 7 = 14).

d)

```
   1,000      74
    978 | 73,251
   (22)    154ˣ
          4,791
             88
           879    remainder
```

```
   1,000       74.89
    978 | 73,251.00
   (22)    154ˣ ˣˣ
          4,791
             88
          8,790
            176
          9,660
            198
            858
```

The answer is 74.9 to one decimal place.

In this problem we had to multiply by 22. This is easy if we remember that 22 is 2 times 11. It is easy to multiply by 2 and by 11 using our shortcut. For instance, we had to multiply 8 times 22. Multiply 8 by 2, then by 11.

$$8 \times 2 = 16$$
$$16 \times 11 = 176$$

To divide by 19, 29 or 39 it would be easier to use our method of direct division, but when the divisor is just below 100, 200, a multiple of 100 or 1,000, you may find this method easier.

You should be able to solve problems like 1,312 divided by 96 in your head. You would divide by 100 minus 4. One hundred divides into 13 hundred thirteen times, so you should be able to say immediately, "Thirteen with 4 times 13 remainder, plus 12, or, 13 with 64 remainder."

Then, if you want the answer to one decimal place, multiply 64 remainder by 10 and divide again. Six hundred and forty divided by 96 is 6 with 40 plus 24 remainder, 64. This is obviously going to repeat so you take the answer to as many decimal places as you like. The answer to three decimal places is 13.667.

To complete this chapter, let us compare this method with the regular method for long division.

Example:

$$705 \div 94 =$$

The "addition" method:

```
100     7
94 | 705
 ⑥    42
      ‾‾
      47
```

How many times does 100 divide into 705? Seven times.

Then we multiply 6 times 7 and add the answer to 5 (in 705) to get our remainder. It is easy to multiply 6 times 7 and easy to add 42 plus 5.

Now, let's compare this with regular long division.

```
        7
      705
94 | 658
    ‾‾‾
      47
```

How many times does 94 divide into 705? Seven times.

Then we multiply 7 times 94 to get our answer of 658. We subtract 658 from 705 to get our remainder.

Our method is much easier, don't you think?

Chapter Sixteen

Checking Answers: Part Two

We can use our strategy of casting out nines for checking answers of division problems.

For example, if we want to check that 42 divided by 2 equals 21, you could cast out the nines. Firstly, we need to generate substitute numbers:

$$42 \div 2 = 21$$
$$6 \div 2 = 3$$

This check is straightforward and needs no special explanation. But how about the following problem?

$$161 \div 7 = 23$$
$$8 \div 7 = 5$$

Here we need to adopt a strategy similar to that which we used for subtraction. An easy way to check division is to multiply the answer by the divisor to give the original number being divided. So we can write the problem thus:

$$8 = 5 \times 7 \quad \text{or} \quad 7 \times 5 = 8$$

Is this correct?

$$5 \times 7 = 35 \quad \text{and} \quad 3 + 5 = 8$$

Yes, our answer is correct.

What if we are checking an answer with a remainder? How would we check the following problem?

$$165 \div 7 = 23 \text{ r}4$$

Have we made a mistake? Casting out nines will find most mistakes without doing the calculation again.

When casting out nines for division problems we rewrite the calculation as a multiplication. But what do we do with the remainder? We write it as:

$$23 \times 7 = 165 - 4$$

We subtract the remainder from the number we divided. That is because we had a remainder of 4 after we divided the number. If the number had been smaller by 4 we would not have had a remainder.

Using substitute numbers, the equation becomes:

$$5 \times 7 = 3 - 4$$

The remainder (4) is greater than the answer (3), so either subtract it first (165 − 4 = 161) or add 9 to the number we are subtracting from. Either way, we end up with:

$$5 \times 7 = 8$$
$$35 = 8$$
$$8 = 8$$

So our answer was correct.

However, there is an important note:

> **You cannot cast out the nines to check a division problem when the answer is rounded off to a specified number of decimal places. Casting nines can only be used to check exact answers.**

Check these problems for yourself by casting nines:

a) $248{,}746 \div 721 = 345 \text{ r}1$ c) $6{,}054 \div 17 = 356 \text{ r}2$

b) $36{,}679 \div 137 = 26{,}722$ d) $3{,}283 \div 49 = 67$

Answers: a), c) and d) were right, b) was wrong.

Casting Out Elevens

Casting out elevens is another simple method for checking answers. It has an advantage over casting out nines in that it will detect if the decimal is wrong by one place or if a zero is added or missing. It is useful if used with casting out nines as a double-check. If I am multiplying by a multiple of 11 (e.g., 66 or 77) I will use this method as well as casting out nines.

I will now show you two easy methods for finding the elevens remainder.

> **The rule is simple with two-digit numbers; subtract the tens digit from the units digit. If the units digit is smaller, add 11 to it.**

For example:

$$13 \times 14 = 182$$

The first number is 13. Subtract 1 (tens digit) from the 3 (units digit) to get an answer of 2. Thus, 2 is the elevens remainder.

The second number, 14 has an elevens remainder of 3 ($4 - 1 = 3$).

To find the elevens remainder of a longer number:

> **Mark off the evenly placed digits going back from the decimal. Subtract the evenly placed digits from the oddly placed digits.**

To find the elevens remainder of 182 we mark the evenly placed digits.

$$1\ 8\ 2$$
$*$

Eight is the second digit back from the decimal. The odd-placed digits are 1 and 2. Adding the odd digits we get:

$$1 + 2 = 3$$

We can't subtract 8 from 3 so we add 11 to the 3.

$$3 + 11 = 14$$
$$14 - 8 = 6$$

Six is our check answer.

Our substitutes for 13 and 14 were 2 and 3. Multiplying them should equal our check answer.

$$2 \times 3 = 6$$

This is the same as our check answer, so our answer is correct.

Casting out nines would have given us the check with less effort. Why cast out elevens? If our answer had been 18.2 instead of 182, casting out nines would not have picked up the mistake but casting out elevens would. Had our answer been 1,712 (incorrectly using our method for multiplying teens), casting out nines would check the answer as correct. But once again, casting out elevens would indicate an error.

Let's check both of these answers by casting out elevens:

$$1\ 8.\ 2$$
$$*\quad\ \ *$$

The evenly placed digits add to 3 $(1 + 2 = 3)$. The oddly placed digit is 8. Our elevens remainder is 5 $(8 - 3 = 5)$.

Our problem becomes:

$$2 \times 3 = 5$$

This is obviously wrong.

Our other incorrect answer was 1,712.

$$1\ 7\ 1\ 2$$
$$*\quad\ *$$

Adding the evenly placed digits $(1 + 1)$ we get 2. The oddly placed digits (7 and 2) add to 9.

$$9 - 2 = 7$$

Our problem, which is also wrong, becomes:

$$2 \times 3 = 7$$

136

Casting out elevens found both answers to be wrong whereas casting out nines did not find an error.

Let's try one more example.

$$1.3 \times 14 = 18.2$$

With 1.3, 1 is the oddly placed digit counting back from the decimal, which makes 3 the evenly placed digit.

We subtract 3 from 1. Because 1 is smaller than 3, we add 11.

$$11 + 1 = 12$$

Now we can subtract 3 from 12.

$$12 - 3 = 9$$

We subtract 1 from 4 to get our elevens remainder for 14.

$$4 - 1 = 3$$

With our answer, 18.2, 1 and 2 are the evenly placed digits and 8 is our odd digit.

$$1 + 2 = 3$$
$$8 - 3 = 5$$

Our check problem becomes:

$$9 \times 3 = 5$$

Nine times 3 is 27. To find the substitute for 27 we subtract 2 from 7.

Seven minus 2 equals 5, the same as our check answer.

An answer of 1.82 or 182 would have checked incorrect by casting out elevens.

Find the elevens remainder for the following numbers:

a) 123

b) 5,237

c) 716

d) 625,174

e) 2,156

f) 8,137

The answers are:

a) 2   b) 1   c) 1 (12, then 1)

d) 0   e) 0   f) 8

If you weren't sure how to find the elevens remainder, go back and read the explanation again. It is worth the effort to learn the method.

Now you need to be able to put the method into practice. Here are some sample problems. Check these answers for yourself:

a) 17 × 17 = 289   c) 32 × 41 = 1,312

b) 154 × 23 = 3,542   d) 46 × 42 = 1,942

One of the answers is incorrect—I won't say which one. Casting out nines would have found the mistake as well. Try this as a double-check.

With either method, casting out nines or casting out elevens, I also like to make a further check by estimating the answer.

These are useful tools if you work with numbers—either at school or on the job.

Chapter Seventeen

Estimating Square Roots

When we square a number, we multiply it by itself. For instance, 4 squared is 16 because 4 times 4 equals 16.

Finding a square root is doing this process in reverse. To find the square root of 16, we have to find which number multiplied by itself will give an answer of 16. The answer, of course, is 4. Likewise, the square root of 25 is 5, because 5 times 5 equals 25.

What is the square root of 64? The answer is 8 because 8 times 8 equals 64.

How about the square root of 56? That is more difficult because it is not an exact square. Seven times 7 is 49, which is too low and 8 times 8 is 64, which is too high. The answer must lie somewhere between 7 and 8. This is how we estimate the square root. We choose the number whose square is just below the number we are working with—in this case, 56—and divide it into the number.

We choose 7, which has a square just below 56. Eight is too high because the 8 squared is above 56.

We now divide 7 into 56 to get an answer of 8.

Then average 7 and 8, or split the difference. The answer is 7.5. (One way of finding the average is to add the numbers and then divide the answer by how many numbers you are averaging.) This answer will always be slightly higher than the correct one so we can round off downwards to 7.48.

The answer, 7.48, is accurate to three digits. Our first answer of 7.5 was accurate to two digits. Often, that is all the accuracy we need.

This is the sign for a square root: $\sqrt{}$

The symbol is written before the number. $\sqrt{16} = 4$ means the square root of 16 equals 4.

Let's try an example:

$$\sqrt{70} =$$

First, try to guess the square root.

$$\sqrt{70} \approx 8 \ (8 \times 8 = 64)$$

Now, divide the number by your approximate answer.

$$70 \div 8 = 8.75$$

Then, split the difference between your estimate (in this case, 8) and the answer you got after dividing the number by your estimate, 8.75. The difference equals:

$$8.75 - 8 = 0.75$$

Splitting the difference (dividing by 2) we get:

$$0.75 \div 2 = 0.375$$

Finally, add this result to the original estimate, 8:

$$8 + 0.375 = 8.375$$

The answer will always be slightly high, so round off downwards. In this case, make the answer 8.37. This answer is accurate to within one-twentieth of 1 percent.

Let's try another. How would we calculate the square root of 29?

$$\sqrt{29} =$$

We choose 5 as our first estimate ($5 \times 5 = 25$). We divide 29 by 5 to get a more accurate estimate.

Five divides into 29 five times with 4 remainder. Five divides into 40 (the 4 remainder times 10) eight times which gives us an answer of 5.8.

$$29 \div 5 = 5.8$$

The difference between 5 and 5.8 is 0.8. Half of 0.8 is 0.4. Add this to the 5 we used as our first estimate to get a corrected estimate of 5.4.

The actual answer is 5.385, but 5.4 is accurate to two digits. We had an error of around 0.2 of 1 percent. That is sufficiently accurate for most purposes.

Let's try another:

$$\sqrt{3,125} =$$

Pair off the numbers backwards from the decimal point:

$$\sqrt{31\ 25} =$$
$$\quad\ \ *\ \ *$$

For each digit pair in the number we are working with we allocate one digit of the answer.

In this example, our answer will be a two-digit number, disregarding any digit after the decimal point.

Even if you have a single digit for the first left-hand digit pair, it still counts as a digit pair.

To calculate the first digit of the answer, we estimate the square root of the first two digits. Our estimate of the square root of 31 is 5 (5 × 5 = 25). The following digits in our estimated answer will always be zeros. There is one more digit in the answer, so we add one zero to the 5 to get 50.

To divide by 50, we divide by 10, and then by 5:

$$3,125 \div 10 = 312.5$$

We now divide by 5 to get 62.5:

$$5 \overline{\left| 312.5 \right.}$$
$$\quad\ \ 62.5$$

Find and split the difference:

$$62.5 - 50 = 12.5$$
$$12.5 \div 2 = 6.25$$

Round off downwards to 6 and add this to your original estimate:

$$50 + 6 = 56$$

$$\sqrt{3{,}125} = 56 \quad \textsf{ANSWER}$$

Using a calculator the answer is:

$$\sqrt{3{,}125} = 55.9$$

Our calculated answer is accurate to within two-tenths of 1 percent. Had we not rounded off downward, our answer still would have been accurate to within 1 percent.

These calculations can easily be done in your head. Yet most people are unable to calculate square roots using a pen and paper.

A Mental Calculation

Do the next calculation in your head.

What is the square root of 500 ($\sqrt{500}$)?

Firstly, we break the number into digit pairs. How many digit pairs do we have? Two. So there are two digits in the answer.

What is the first digit pair? It only has one digit, 5. What is the square root of 5? We choose 2, because $2 \times 2 = 4$.

We make up the second digit of the square root with a zero. Our estimate is 20.

We now have to divide 500 by 20. How do we do this? We divide 500 by 10, then by 2.

$$500 \div 10 = 50$$
$$50 \div 2 = 25$$

We now split the difference between 20 and 25 to get 22.5. We round downwards to 22.4.

The answer, using a calculator, is 22.36.

Our estimate of 22.5 had about half of 1 percent error. Our rounded off estimate of 22.4 had less than 0.2 of 1 percent error. That's pretty good for a mental calculation—especially when we realize that the only method most people know for calculating square roots is to use a calculator. Calculating square roots in your head will almost certainly gain you a reputation for being a mathematical genius.

Let's try another example:

$\sqrt{93,560} =$

We pair off the digits.

$\sqrt{9\ 35\ 60} =$
$\quad *\quad *\quad *$

The first digit pair is 9. The square root of 9 is 3 (3 × 3 = 9). There are three digit pairs, so we add two zeros to the 3 to make up three digits. Our estimate is 300.

To divide by 300, we divide by 100 and then by 3. (To divide by 100, move the decimal two places to the left.)

93,560 ÷ 100 = 935.60

3 | 935.60
$\quad\overline{}$
$\quad$ 311.86

311.86 − 300 = 11.86

11.86 ÷ 2 = 5.93, rounded down to 5.9

300 + 5.9 = 305.9

With a calculator we get an answer of 305.8758. We have an error of 0.0079 percent.

Let's try one more together:

$\sqrt{38,472,148} =$

That looks like an impressive problem. If we were doing this in our head, we could round off downwards before we begin. Let's do the full calculation first.

Firstly, we pair off the digits:

$$\sqrt{38\ 47\ 21\ 48} =$$
$$* \quad * \quad * \quad *$$

There are four digit pairs so there will be four digits in the answer.

The first digit pair is 38. We estimate the square root of 38 as 6 because 6 times 6 is 36. We make up the other digits with zeros. Our estimate is 6,000.

We divide 38,472,148 by our estimate. First divide by 1,000 and then by 6.

$$38,472,148 \div 1,000 = 38,472.148$$

Because we are only estimating, we can drop the digits after the decimal. We now divide 38,472 by 6:

$$38,472 \div 6 = 6,412$$

We split the difference between 6,000 and 6,412. The difference is 412, and half of 412 is 206. (Half of 400 is 200 and half of 12 is 6.)

Add 206 to our first estimate and we get 6,206. We round off downwards to get:

6,200   ANSWER

The actual answer is 6,202.59. For practical purposes, our estimate was probably close enough. However, if we want an accurate answer, the method taught in the next chapter is the easiest method I know.

In the meantime, calculate these for yourself. See how many you can do without using pen and paper.

a) $\sqrt{1,723} =$ e) $\sqrt{5,132} =$

b) $\sqrt{2,600} =$ f) $\sqrt{950} =$

c) $\sqrt{80} =$ g) $\sqrt{2,916} =$

d) $\sqrt{42} =$ h) $\sqrt{1,225} =$

The answers are:

a) 41.5   b) 50.99   c) 8.94   d) 6.48

e) 71.64   f) 30.82   g) 54   h) 35

When the Number Is Just Below a Square

The more accurate your estimate, the more accurate your final answer will be. Thus, we need to choose an estimate as close as possible to the actual square root.

The examples we have worked with thus far are just above a square. For instance, in the exercises you just tried, 2,600 is just above 50 squared (2500), so we use 50 as our estimate.

Here is a procedure for working with numbers that are just below a square. To get a more accurate answer, instead of choosing the lower number as our first estimate, we can choose the higher number if it is closer to our answer.

For example:

$$\sqrt{2,400} =$$

Split the number into digit pairs:

$$\sqrt{24\ 00} =$$
$$\quad * \quad *$$

We estimate the square root of 24 as 5 because 24 is closer to 5 squared (25) than it is to 4 squared (16). Our estimate of the square root of 2,400 is 50.

We now divide 2,400 by 50. To divide by 50, divide by 100, then double the answer (50 = 100 ÷ 2).

$$2,400 \div 100 = 24$$
$$24 \times 2 = 48$$

Split the difference between 48 and 50.

$$50 - 48 = 2$$
$$2 \div 2 = 1$$

Adding 1 to 48 gives our answer: 49.

A calculator gives the answer as 48.98979. We had an error of about 0.02 of 1 percent.

Let's try another:

$$\sqrt{6,300} =$$

We pair the digits:

$$\sqrt{63\ 00} =$$
$$^{*}\qquad ^{*}$$

Our estimate of the first digit pair is 8 because 63 is much closer to 8 squared (64) than it is to 7 squared (49). Thus, our estimate is 80.

We divide by 10, then by 8:

$$6,300 \div 10 = 630$$
$$630 \div 8 = 78.75$$

We now split 78.75 and 80. We can subtract 78.75 from 80, halve the answer, and minus the result from 80.

Here is the good news—we have a shortcut.

A Shortcut

What we are really doing is averaging the amounts.

To average 78.75 and 80, we add them together (158.75) and halve the result.

Here is the shortcut. We know the answer is in the seventies, so 7 is the first digit of the answer. For the rest of the answer, put a 1 in front of the 8.75 (to get 18.75) and halve it. No further addition or subtraction is needed.

Half of 18 is 9. Place the nine next to the 7 to get 79. Half of 75 is 37.5. Our answer becomes 79.375. We would round off downwards to 79.37.

The actual answer is 79.3725, which means we had an error of 0.003 of 1 percent. Had we used 70 as our first estimate we would have got an answer of 80.

Why did our shortcut work? Because, to average the numbers, 78.75 and 80, we add them and then halve the answer.

$$78.75 + 80 = 158.75$$
$$158.75 \div 2 = 79.375$$

When we divided 15 by 2 we got an answer of 7 and carried 1 to the 8 to make 18. In our shortcut we simply skipped that part of the calculation.

Greater Accuracy

If we need greater accuracy for our calculation we can repeat the procedure using our estimate answer as a second estimate.

Let's use the first example cited in this chapter:

$$\sqrt{56} =$$

Our first estimate is 7 ($7 \times 7 = 49$).

$$56 \div 7 = 8$$
$$8 - 7 = 1 \text{ (the difference)}$$
$$1 \div 2 = 0.5$$
$$7 + 0.5 = 7.5$$

Now, repeat the process. Divide 56 by 7.5. It isn't difficult to divide by 7.5. It is the same as $112 \div 15$ or $224 \div 30$. If we double both numbers the ratio stays the same.

Two hundred and twenty-four is easily divided by 30. We divide by 10 (22.4) and then by 3.

$$224 \div 30 = 7.4667$$

We can use our shortcut to average the answers. We know the first part of the answer is 7.4. We put our carried 1 in front of 667 to get 1,667 and divide by 2.

$$1,667 \div 2 = 833.5$$

We attach 833 to 7.4 for our answer of 7.4833. This answer is correct for all expressed digits.

147

Generally speaking, each time we repeat this process we double the number of significant digits which are accurate.

Let's try another.

One of our earlier mental calculations was the square root of 500. We can continue the process in our head to increase the accuracy of our answer.

We calculated $\sqrt{500}$ = 22.5.

Instead of dividing 500 by 20, we now divide by 22.5. Is this difficult? No, not if we double our numbers twice.

500 and 22.5 doubled is 1,000 and 45. Doubled again we get 2,000 and 90.

We divide 2,000 by 90 to get a more accurate estimate. To divide 2,000 by 90 we divide by 10 and then by 9.

$$2,000 \div 10 = 200$$
$$200 \div 9 = 22.22$$

We now split the difference between 22.22 and 22.5.

The 22 in front of the decimal point is obviously correct. To find out what comes after it we simply average 22 and 50.

$$22 + 50 = 72$$
$$72 \div 2 = 36$$

Add 0.36 to 22 for an answer which is accurate for all digits given.

$$22 + 0.36 = 22.36 \quad \text{ANSWER}$$

With a little practice, this procedure can be carried out entirely in your head. Try it!

Chapter Eighteen

Calculating Square Roots

There is an easy method for calculating the exact answer for square roots. It involves a process that I call cross multiplication.

Here's how it works.

Cross Multiplication

To cross multiply a single digit, you square it.

$$3^2 = 3 \times 3 = 9$$

If you have two digits in a number, you multiply them and double the answer. For example:

$$34 = 3 \times 4 = 12$$
$$12 \times 2 = 24$$

With three digits, multiply the first and third digits, double the answer, and add this to the square of the middle digit. For example, 345 cross multiplied is:

$$3 \times 5 = 15$$
$$15 \times 2 = 30$$
$$30 + 4^2 = 46$$

The general rule for cross multiplication of an even number of digits is:

Multiply the first digit by the last digit, the second by the second last, the third by the third last and so on, until you have multiplied all of the digits. Add them together and double the total.

In practice, you would add them as you go and double your final answer.

The general rule for cross multiplication of an odd number of digits is:

Multiply the first digit by the last digit, the second by the second last, the third by the third last and so on, until you have multiplied all of the digits up to the middle digit. Add the answers and double the total. Then square the middle digit and add it to the total.

Here are some examples:

$$123 = 1 \times 3 = 3, 3 \times 2 = 6, 6 + 2^2 (4) = 10$$
$$1{,}234 = 1 \times 4 \ (4), + 2 \times 3 \ (6) = 10, 10 \times 2 = 20$$
$$12{,}345 = 1 \times 5 \ (5), + 2 \times 4 \ (8) = 13, 13 \times 2 = 26, 26 + 3^2 \ (9) = 35$$

Using Cross Multiplication to Extract Square Roots

Here is the method for extracting square roots.

For example:

$$\sqrt{2{,}809} =$$

Firstly, pair the digits back from the decimal. There will be one digit in the answer for each digit pair in the number.

$$\sqrt{28\ 09} =$$
$$\quad * \quad *$$

This problem will have a two-digit answer.

Secondly, estimate the square root of the first digit pair. The square root of 28 is 5 ($5 \times 5 = 25$). So 5 is the first digit of the answer.

Double the first digit of the answer (2 × 5 = 10) and write it to the left of the number. This number will be our divisor. Write 5, the first digit of our answer, above the 8 in the first digit pair, 28.

Your work should look like this:

$$\begin{array}{r} 5 \\ 10 \ \ \sqrt{28\ 09} = \end{array}$$

We have finished working with the first digit of the answer.

To find the second digit of the answer, square the first digit of your answer and subtract the answer from your first digit pair.

$5^2 = 25$
$28 - 25 = 3$

Three is our remainder. Carry the 3 remainder to the next digit of the number being squared. This gives us a new working number of 30.

Divide our new working number, 30, by our divisor, 10. This gives 3, the next digit of our answer. Ten divides evenly into 30, so there is no remainder to carry. Nine is our new working number.

Our work should look like this:

$$\begin{array}{r} 5 \ \ 3 \\ 10 \ \ \sqrt{28 \ _3 09} = \\ 25 \end{array}$$

Finally, cross multiply the last digit of the answer. (We don't cross multiply the first digit of our answer. After the initial workings the first digit of the answer takes no further part in the calculation.)

$3^2 = 9$

Subtract this answer from our working number.

$9 - 9 = 0$

There is no remainder: 2,809 is a perfect square. The square root is 53.

$10 \ \ \sqrt{2,809} = 53$

Let's try another example:

$$\sqrt{54,756} =$$

Firstly, pair the digits. There are three digit pairs. The answer will be a three-digit number.

$$\sqrt{5\ 47\ 56} =$$
$$*\quad*\quad*$$

Now estimate the square root of the first digit pair. Five is the first digit pair. Our estimate is 2 ($2 \times 2 = 4$).

Write 2 as the first digit of the answer. Double the first digit of the answer to get our divisor ($2 \times 2 = 4$).

Our work looks like this:

$$\begin{array}{r} 2 \\ \hline 4\ \sqrt{5\ 47\ 56} \\ 4 \end{array}$$

Square the first digit of the answer, write it underneath and subtract this from the first digit pair.

$$2^2 = 4$$
$$5 - 4 = 1$$

We carry the 1 to the next digit. We now have a working number of 14.

Divide 14 by our divisor, 4. The answer is 3 with a remainder of 2 ($3 \times 4 = 12$). Carry the remainder to the next digit. Our next working number is 27.

$$\begin{array}{r} 2\ \ 3 \\ \hline 4\ \sqrt{5_14_27\ 56} \\ 4 \end{array}$$

Cross multiply. We don't include the first digit of the answer so we cross multiply the 3.

$$3^2 = 9$$

Subtract the cross multiplication from the working number.

$$27 - 9 = 18$$

Divide 18 by 4 for an answer of 4 with 2 remainder. So 4 is the final digit of the answer. Any other digits we work out now go to the right of the decimal point. We carry the 2 remainder.

$$
\begin{array}{r}
2\ 3\ 4 \\
4\ \ \sqrt{5{,}_1 4{,}_2 7{,}_2 5\ 6} \\
4
\end{array}
$$

We now have a working number of 25.

Cross multiply.

$$4 \times 3 = 12$$
$$12 \times 2 = 24$$

Subtract 24 from our working number, 25 for an answer of 1. Divide 1 by 4. We get an answer of zero with 1 remainder. Carry the 1 to the final digit of the number. We now have a working number of 16.

$$
\begin{array}{r}
2\ 3\ 4.0 \\
4\ \ \sqrt{5{,}_1 4{,}_2 7{,}_2 5{,}_1 6} \\
4
\end{array}
$$

Cross multiply:

$$0 \times 3 = 0$$
$$4^2 = 16$$

Subtract 16 from our working number to get an answer of zero. We have no remainder.

Again, 54,756 is a perfect square. The square root is 234.

If we had a remainder, we would simply carry it to the next digit and continue the calculation to however many decimal places we wish.

Comparing Methods

What would our answer have been had we estimated the square root by the method taught in the previous chapter?

$$\sqrt{5\ 47\ 56} =$$
$$* \quad * \quad *$$

We estimate 2 as the first digit of the answer. The following digits automatically become zeros. Our first estimate is 200.

Divide 54,756 by 200. We divide by 100, then by 2.

$$54{,}756 \div 100 = 547.56$$
$$547 \div 2 = 273$$

Splitting the difference, we get 236. We would round off downwards to 235—one off the actual answer, or an error of about one half of a percent. This is probably acceptable for most purposes. But if you want an accurate calculation, the cross multiplication strategy is the easiest method I am aware of to calculate square roots.

Try these for yourself.

a) $\sqrt{3{,}249} =$

b) $\sqrt{2{,}116} =$

c) $\sqrt{103{,}041} =$

The answers are:

a) 57   b) 46

Let's work through c) together:

$$\sqrt{103{,}041} =$$

We split the number into digit pairs.

$$\sqrt{10\ 30\ 41} =$$
$$* \quad * \quad *$$

There are three digit pairs so we know there will be three digits in the answer before the decimal point.

We estimate the square root of the first digit pair. The first digit pair is 10. Three times 3 is 9. Four squared would be too high. The first digit of our answer is 3. Therefore, our divisor is 6.

Three squared is 9. Subtracting 9 from 10 gives us a remainder of 1. Carry the remainder to the next digit. This gives us a new working number of 13.

$$\begin{array}{r} 3 \quad\quad\quad \\ 6 \ \sqrt{10_{\,1}30\ 41} \\ 9 \quad\quad\quad\quad \end{array}$$

Divide 13 by our divisor, 6.

$$13 \div 6 = 2 \text{ r}1$$

The next digit of our answer is 2. Our next working number is 10.

$$\begin{array}{r} 3\ 2 \quad\quad\quad \\ 6 \ \sqrt{10_{\,1}3_{\,1}041} \\ 9 \quad\quad\quad\quad \end{array}$$

We cross multiply the 2 to get 4. Subtract 4 from our working number.

$$10 - 4 = 6$$

Divide 6 by 6.

$$6 \div 6 = 1$$

One is the final digit of our answer. There is no remainder to carry.

$$\begin{array}{r} 3\ 2\ 1 \quad\quad\quad \\ 6 \ \sqrt{10_{\,1}3_{\,1}0_{\,0}4\ 1} \\ 9 \quad\quad\quad\quad \end{array}$$

Our new working number is 4. We cross multiply. Twenty-one cross multiplied is 4 ($2 \times 1 = 2$, $2 \times 2 = 4$). Subtract 4 from 4 for an answer of zero.

$$\begin{array}{r} 3\ 21.0 \quad\quad\quad \\ 6 \ \sqrt{10_{\,1}3_{\,1}0_{\,0}4_{\,0}1} \\ 9 \quad\quad\quad\quad \end{array}$$

Our new working number is 1.

Cross multiply.

> $0 \times 2 = 0$
> $1^2 = 1$

Subtract 1 from 1. Our final answer is zero, so 103,041 is a perfect square. The square root is exactly 321.

Also, with a little practice, you should be able to perform the above calculation entirely in your head. That is impressive.

A Question from a Reader

A reader asked me how I would find the square root of 2,401.

Paired off, the problem looks like this:

> $\sqrt{24\ 01}$
>   *   *

We have two digit pairs so there are two digits in the answer.

He says, "I get the 4 as my estimate of the square root of 24 ($4 \times 4 = 16$) and a divisor of 8, but 16 from 24 leaves 8, and 8 into 80 goes 10 times, so what am I doing wrong?"

Nothing! Because 10 is not a valid digit, we reduce it by 1 to get a second digit of the answer of 9 with a remainder of 8 which you carry to the 1, making 81.

Cross multiply the 9 (9 squared) = 81. Subtract 81 from the next working number, 81.

> $81 - 81 = 0$

So we have zero remainder. The answer, 49, is exact.

He then asked how I would calculate the following square root:

> $\sqrt{23{,}222{,}761}$

Paired off, the problem looks like this:

> $\sqrt{23\ 22\ 27\ 61}$
>   *   *   *   *

The fully worked problem would look like this:

$$
\begin{array}{llllllll}
 & 4 & 8 & 1 & 9. & 0 & 0 & 0 \\
8\ \sqrt{23} & {}_7 2 & {}_8 2 & {}_{10}2 & {}_{14}7 & {}_2 6 & {}_8 1 \\
-16 & & -64 & -16 & -145 & -18 & -81 \\
 & & 18 & 86 & 2 & 8 \\
\end{array}
$$

Each time you divide you have to keep in mind the remainder carried must give a total higher than the cross multiplication. Also, 9 is the highest possible digit, even if the divisor will go 10 or 11 times. If the divisor divides exactly into the number leaving no remainder and you already have digits in the answer to cross multiply, you must drop the answer by at least 1.

I usually use the first method to estimate a square root, but if I need an exact answer, this is the way to go. The larger the number you are working with, the more difficult the calculation as you have to cross multiply larger numbers as you go. Then the estimation method is easier.

Let's compare both methods for finding the square root of 196. Firstly, by estimation:

$$\sqrt{1\ 96} =$$
$$*\quad *$$

We split the number into digit pairs. There are two digit pairs, so we have two digits in the answer.

We estimate the square root of the first digit pair, 1.

The square root of 1 is 1. We have the first digit of the answer, 1.

We supply a zero for the next digit. Our first estimate is 10.

$$196 \div 10 = 19.6$$

We round off downwards and split the difference. Round off to 19. The difference is 9 ($19 - 10 = 9$). Half of 9 is 4.5. We add 4.5 to 10 to get our answer of 14.5.

For increased accuracy, we could use 15 as a second estimate.

Divide 196 by 15. The easy way to do that is to double both numbers (196 = 200 − 4, double 200 − 4 is 400 − 8). We now have 392 ÷ 30. To divide 392 by 30, we divide by 10 and then by 3.

$$392 ÷ 10 = 39.2$$
$$39.2 ÷ 3 = 13.06$$

We round this off to 13.

Split the difference between 13 and 15. Half of 2 is 1. Minus this from our originally rounded-up estimate (15 − 1 = 14). The answer is 14, which happens to be the correct answer.

How would our second method using cross multiplication work for calculating the square root of 196?

$$\sqrt{196} =$$

Split the number into digit pairs:

$$\sqrt{1\ 96} =$$
$$*\quad *$$

Estimate the square root of the first digit pair. The square root of 1 is 1. That is the first digit of the answer. We double the first digit, 1, to find our divisor. Two times 1 is 2.

$$1$$
$$2\ \sqrt{1\ 96}$$

We divide 2 into the next digit, 9.

$$9 ÷ 2 = 4\ and\ 1r$$

Write 4 as the second digit of the answer and carry the 1 remainder to the next digit, 6, making 16.

$$1\ 4.0$$
$$2\ \sqrt{1\ 9\,_16}$$

Cross multiply (square) the 4 of our answer and subtract from our working number of 16. Four squared is 16, subtracted from 16 gives us zero remainder. 196 is a perfect square.

In this particular case, the method taught in this chapter is the easier method for calculating the square root. You now have a choice of methods for finding square roots of numbers.

Chapter Nineteen

Fun
Shortcuts

Many books have been written about mathematical shortcuts. Such shortcuts will not only save time and effort, they can be useful in developing number skills as well. In this chapter, I will show you several fun shortcuts.

Multiplication by 11

To multiply a two-digit number by 11, simply add the two digits together and insert the result in between.

For example, to multiply 23 by 11, add 2 plus 3 , which equals 5, and insert the 5 between the 2 and the 3. The answer is 253.

To multiply 14 by 11, add 1 plus 4, which equals 5, and insert the 5 between the 1 and the 4. The answer is 154.

Try these for yourself:

a) $63 \times 11 =$	d) $26 \times 11 =$
b) $52 \times 11 =$	e) $71 \times 11 =$
c) $34 \times 11 =$	f) $30 \times 11 =$

The answers are:

a) 693	b) 572	c) 374
d) 286	e) 781	f) 330

In the examples above, the two digits add up to less than 9. What do we do if the two digits add to a number higher than 9? When the result is a two-digit number, insert the units digit of the result between the digits and carry the one to the first digit of the answer.

For instance, to multiply 28 by 11, add 2 to 8, which equals 10. Insert the 0 between the 2 and 8 to get 208, and carry the 1 to the first digit, 2, to give an answer of 308.

Let's try 88 × 11:

$$8 + 8 = 16$$

Insert 6 between the 88 to give 868, then add 1 to the first 8 to give an answer of 968.

Calling Out the Answers

Giving children these problems (and doing them ourselves) will develop basic number skills. Children of all ages will become very quick at calling out the answers.

If someone were to ask you to multiply 77 by 11, you would immediately see that 7 plus 7 equals 14, which is more than 9. You would immediately add 1 to the first 7 and call out, "Eight hundred and . . ." The next digit will be the 4 from the 14, followed by the remaining 7, so then you could say, ". . . forty . . . seven." Try it. It is much easier than it sounds.

Here is another example: if you had to multiply 84 by 11, you would see 8 plus 4 is more than 9, so you would add 1 to the 8 and say, "Nine hundred and . . ." Now we add 8 and 4, which is 12, so the middle digit is 2. You would say, ". . . twenty . . ." The final digit remains 4, ". . . four." Your answer is, "Nine hundred and twenty-four."

How about 96 × 11?

Nine plus 6 is 15. Add one to the 9 to get 10. Work with 10 as you would with a single-digit number: 10 is the first part of the answer. Five is the middle digit and 6 remains the final digit. The answer is 1,056.

If you were doing this problem in your head, you would visualize, "Nine plus one carried is ten." Out loud you would say, "One thousand and . . ." You would then see that the 5 from the 15 is the tens digit, so you continue, ". . . fifty . . ." The units digit remains the same, 6. You can then give the answer, "One thousand and . . . fifty . . . six." Or you could say, "Ten . . . fifty-six."

Try these for yourself. Call out the answers as fast as you can.

a) 37 × 11 =   d) 92 × 11 =

b) 48 × 11 =   e) 82 × 11 =

c) 76 × 11 =   f) 66 × 11 =

The answers are:

a) 407   b) 528   c) 836

d) 1,012   e) 902   f) 726

Multiplying Multiples of 11

How would you multiply 330 by 12?

It doesn't seem like we can use our shortcut for multiplication by 11, but have another look.

330 = 3 × 11 × 10

(Get used to ignoring the zero on the end of a number when you have to use it in a multiplication or division. It is just a statement that the number in front has been multiplied by ten.)

Because 33 is three elevens, we can multiply 12 by 3, then by 11. Twelve times 3 is 36, and 36 times 11 is 396—using our shortcut. Multiply by 10 and we have 3,960.

Always look to see if you can use this shortcut when multiplying by any multiple of 11; that is, 22, 33, 44, 55, or 2.2, 3.3, 5.5, etc. For example, there are 2.2 pounds in a kilogram. To convert kilograms to pounds you multiply by 2.2. Double the number, multiply by 11, and then divide by ten to account for the decimal place.

To convert 80 kilograms to pounds, double 80 to get 160. Then multiply this by 11 (1,760), and divide by 10 to get 176 pounds.

What is 90 kilograms in pounds?

$90 \times 2 = 180$

$180 \times 11 = 1{,}980$

$1{,}980 \div 10 = 198$ pounds   ANSWER

Multiplying Larger Numbers

To multiply larger numbers by 11, we use a similar method. Let's take the example of $12{,}345 \times 11$. We would write the problem like this:

$012{,}345 \times 11 =$

We write a zero in front of the number we are multiplying. You will see why in a moment. Beginning with the units digit, add each digit to the digit on its immediate right. In this case, add 5 to the digit on its right. There is no digit on its right, so add nothing:

$5 + 0 = 5$

Write 5 as the last digit of your answer. Your calculation should look like this:

$$\frac{012{,}345 \times 11}{5}$$

Now go to the 4. Five is the digit on the right of the 4:

$4 + 5 = 9$

Write 9 as the next digit of your answer. Your calculation should now look like this:

$$\frac{012{,}345 \times 11}{95}$$

Continue the same way:

$3 + 4 = 7$

$2 + 3 = 5$

$1 + 2 = 3$

$0 + 1 = 1$

Here is the finished calculation:

$$\frac{012,345 \times 11}{135,795}$$

If we hadn't written the zero in front of the number to be multiplied, we might have forgotten the final step.

This is an easy method to multiply by 11. The strategy also develops addition skills while students are using the method as a shortcut.

Let's try another problem. This time we will have to carry digits. Note that the only digit you can carry, using this method, is 1.

Let's try this example:

$$217,475 \times 11 =$$

We write the problem like this:

$$0217,475 \times 11$$

We add the units digit to the digit on its right. There is no digit to the right, so 5 plus nothing is 5. Write down 5 below the 5. Now add the 7 and the 5:

$$7 + 5 = 12$$

Write the 2 as the next digit of the answer and carry the 1. Your work should look like this:

$$\frac{02174,75 \times 11}{25}$$

The next steps are:

$$4 + 7 + 1 \text{ (carried)} = 12$$

Write 2 and carry 1. Add the next numbers:

$$7 + 4 + 1 \text{ (carried)} = 12$$

Two is again the next digit of the answer. Carry the 1.

$$1 + 7 + 1 \text{ (carried)} = 9$$

$$2 + 1 = 3$$
$$0 + 2 = 2$$

Here is the finished calculation:

$$\frac{0\,2_1{,}1_1{,}7_1{,}47\,5 \times 11}{2\,3\,9\,2\,22\,5}$$

A Math Game

We can also use this method as a game. This involves a simple check for multiplication by 11. Remember, the problem isn't completed until we have checked it. Let's check our first problem:

$$\frac{012{,}345 \times 11}{135{,}795}$$

Write a cross under every second digit of the answer, beginning from the right-hand end of the number. The calculation will now look like this:

$$\frac{0\ 1\ 2\ 3\ 4\ 5\ \times\ 11}{1\ 3\ 5\ 7\ 9\ 5}$$
$$x\ \ \ x\ \ \ x$$

Now, add the digits with the cross under them:

$$1 + 5 + 9 = 15$$

Fifteen is our check answer. Then, add the digits without the cross:

$$3 + 7 + 5 = 15$$

Fifteen is our second check answer.

If the original answer is correct, the check answers will either be the same, have a difference of 11, or a difference of a multiple of 11, such as 22, 33, 44 or 55. In the example above, both check answers added to 15, so our answer is correct. This is also a good test to see if a number can be evenly divided by 11.

Let's check the second problem.

$$0\ 2\ 1\ 7\ 4\ 7\ 5 \times 11$$
$$\overline{2\ 3\ 9\ 2\ 2\ 2\ 5}$$
$$xxx$$

Add the digits with the cross under them:

$$3 + 2 + 2 = 7$$

Add the digits without the cross:

$$2 + 9 + 2 + 5 = 18$$

To find the difference between 7 and 18, we take the smaller number from the larger number:

$$18 - 7 = 11$$

If the difference is 0, 11, 22, 33, 44, 55, 66, etc., then the answer is correct. Here, we have a difference of 11, so our answer is correct.

Give these problems to children. Ask them to make up their own numbers to multiply by 11 and see how big a difference they can get. The larger the number they are multiplying, the greater the difference. Let them try for a new record.

Children will multiply a one-hundred-digit number by 11 in their attempt to set a new record. While they are trying for a new world record, they are improving their basic addition skills and checking their work as they do so.

Multiplication by 9

Just as there is an easy shortcut for multiplication by 11 because 11 is one more than 10, so there is an easy shortcut for multiplication by 9 because 9 is one less than 10. Instead of adding each digit to the digit on the right, we subtract each digit from the digit on the right.

Because subtraction involves borrowing and carrying, we can make the following shortcut. We subtract the units digit from 10, then subtract each successive digit from 9 and add the neighbor. We subtract 1 from the first digit of the number for the first digit of the answer.

For example:

$254 \times 9 =$

Subtracting 4 from 10 gives us 6. Subtract 5 from 9 is 4, plus 4 is 8 (86). Subtract 2 from 9 is 7, plus 5 is 12. Write 2, carry 1 (286).

Now, subtract 1 from the first digit, 2, and add the 1 carried for our answer, 2 (2,286).

$254 \times 9 = 2^{1}286$ ANSWER

Division by 9

Here is an easy way to divide any two-digit number by 9.

To divide 42 by 9, take the tens digit, 4, as the answer, and add both digits of the number for the remainder. $4 + 2 = 6$ remainder.

4 r6   ANSWER

Let's try another.

$34 \div 9 =$

The tens digit is 3 so 3 is our answer.

$3 + 4 = 7$

3 r7   ANSWER

How about 71?

The tens digit is . . . ?

The remainder is . . . ?

The answer is 7 with 8 remainder.

Try these for yourself.

a) $52 \div 9 =$ c) $61 \div 9 =$

b) $33 \div 9 =$ d) $44 \div 9 =$

The answers are:

a) 5 r7   b) 3 r6   c) 6 r7   d) 4 r8

Those were easy. But you are probably asking yourself, what if you want to divide 46 by 9? We have a problem because 4 plus 6 equals 10. What do we do?

Let's do it.

$$46 \div 9 =$$

The tens digit is 4 so our answer is 4.

Four plus 6 equals 10 remainder. Our remainder is more than 9 so it can't be right. We can't have a remainder that is larger than the divisor. Ten divided by 9 is 1 with 1 remainder. (Our tens digit is 1 plus 1 + 0 = 1 remainder.)

Add the answer of 1 to the previous answer, 4, to get a final answer of 5 with 1 remainder which we got from the second step.

Let's try another.

$$75 \div 9 =$$

The tens digit is 7.

Seven plus 5 equals 12 remainder. We can't have a remainder that is larger than the divisor so we divide the remainder.

$$12 \div 9 =$$

The tens digit is 1. This increases our first answer by 1.

$$7 + 1 = 8$$

$$1 + 2 = 3 \text{ remainder}$$

8 r3   ANSWER

Try these for yourself.

a) $85 \div 9 =$ c) $28 \div 9 =$

b) $37 \div 9 =$ d) $57 \div 9 =$

The answers are

a) 9 r4   b) 4 r1   c) 3 r1   d) 6 r3

Multiplication Using Factors

Often, when you have to multiply two numbers, the calculation can be made easier if you can double one number and halve the other. (The method is sometimes called "double and halve.") What you are really doing is multiplying using factors.

An easy way to multiply 3 times 14 is to double 3 and halve 14 to get 6 times 7. Why does this work? Because, if you break 14 into 2 times 7, you can multiply 3 by 2 and then by 7.

$$3 \times 14 = 3 \times (2 \times 7) = 6 \times 7 = 42$$

To multiply 4 times 22, you would multiply 4 by 2 to get 8, then by 11. (You doubled the 4 and halved 22.) Actually, you just used factors of 22 (2 times 11) to make the calculation easy. Whenever you have to multiply a large number by a smaller number, look to see if you can use this principle to make the calculation easier.

Let's say you need to multiply 14 times 24. You could use 10 as a reference number and the calculation is straightforward.

$$+ \textcircled{4} + \textcircled{14}$$
$$\textcircled{10} \quad 14 \times 24 =$$

Adding crossways we get 28 (14 + 14 or 24 + 4). Multiply 28 by the reference number, 10, to get 280.

Now we have to multiply the numbers in the circles, 4 times 14. We could say 4 times 10 is 40, plus 4 times 4, 16, makes 56.

Or, we could double and halve.

Fourteen is 2 times 7 and 4 times 2 is 8. Hence, 4 times 14 is the same as 8 times 7. Eight times 7 is 56.

We doubled 4 and halved 14 to get 8 times 7.

Our subtotal, 280, plus 56, gives an answer of 336.

Try these:

a) $4 \times 18 =$

c) $48 \times 180 =$

b) $6 \times 24 =$

The answers are:

   a) 72   b) 144   c) 8,640

(The problems became 8×9, 12×12 and 96×90.)

With a little practice you will easily recognize these opportunities.

Division by Factors

If you have a 100-mg bottle of medicine and you have to take two doses each day of 7.5 mg, how many days will the medicine last?

It seems difficult to divide 100 by 7.5.

You don't have to. If you take two doses each day, you are taking 15 mg per day. But you can't divide 100 evenly by 15.

There is an easier way to solve this problem. If we double both values, the answer remains the same. Double 100 divided by 15 is 200 divided by 30.

To divide by 30, we divide by 10, then by 3.

$$200 \div 10 = 20$$
$$20 \div 3 = 6\tfrac{2}{3}$$

The medicine will last for 6 and a half days. (You would take two-thirds of a dose for your last dose to finish the bottle.)

This seems like an impressive calculation but it is really simple.

An easy way to:

⇨   divide by 15 is to double the number and divide by 30.

⇨   divide by 25 is to double the number and divide by 50.

⇨   divide by 35 is to double the number and divide by 70.

⇨   divide by 45 is to double the number and divide by 90.

For example, if you wanted to divide 2,341 by 35, you would double 2,341 and divide by 10, then by 7.

$$2,341 \times 2 = 4,682$$

$$4,682 \div 10 = 468.2$$
$$468.2 \div 7 = 66.8857$$

This is an easy calculation using simple division.

Try these for yourself:

a) $600 \div 15 =$ c) $560 \div 35 =$

b) $217 \div 35 =$ d) $630 \div 45 =$

The answers are:

a) 40   b) 6.2   c) 16   d) 14

Multiplying Two Numbers Which Have the Same Tens Digits and Whose Units Digits Add to 10

In Chapter Ten we learned an easy method for squaring numbers ending in 5. There is a related shortcut that uses a similar formula.

For example, if you want to multiply 13×17, you observe the tens digits are the same and the units digits add to 10.

Firstly, multiply the tens digit by one more than itself.

Adding 1 to the tens digit we get $1 + 1 = 2$. Multiply 1 by 2 to get 2. This will be the hundreds digit (200).

Next, multiply the units digits. Three times 7 equals 21.

$$200 + 21 = 221$$

Another example:

$$62 \times 68 =$$

The tens digits of both numbers is 6. Add 1 to 6 ($6 + 1 = 7$). Multiply 6 by 7 to get 42. This is the number of hundreds so our subtotal is 4,200. Then, $2 \times 8 = 16$.

$$4,200 + 16 = 4,216 \quad \text{ANSWER}$$

If you work with numbers you will find you encounter these situations more often than you would expect.

171

Let's try another:

$$123 \times 127 =$$
$$12 + 1 = 13$$
$$12 \times 13 = 156$$

This is the number of hundreds so we say 15,600.

$$3 \times 7 = 21$$

The answer is 15,621.

Try these for yourself:

a) $43 \times 47 =$ d) $32 \times 38 =$

b) $21 \times 29 =$ e) $46 \times 44 =$

c) $114 \times 116 =$ f) $148 \times 142 =$

The answers are:

a) 2,021   b) 609   c) 13,224

d) 1,216   e) 2,024   f) 21,016

Those were so easy there was almost no effort involved at all. Yet it seems you're performing calculations like a genius. This only goes to show, geniuses generally know better methods than the rest of us. Learn their methods and you will perform like a genius.

Multiplying Numbers When the Units Digits Add to 10 and the Tens Digits Differ by 1

If you had to multiply 38 by 42 there is an easy shortcut you can use.

When the units digits add to ten and the tens digits differ by one, the lower number is the same amount lower than the tens digit as the other is higher. In this case, 38 is 2 lower than 40 and 42 is 2 higher. There is a rule in mathematics that if you multiply two numbers that differ above and below a square by the same amount, the answer will be that number squared less the difference squared.

Let's take our example:

38 × 42 =

Thirty-eight is 2 lower than 40 and 42 is 2 higher. It is easy to multiply 40 × 40: 40 squared is 1,600. (To multiply 40 times 40, drop the zeros, multiply 4 times 4 is 16, then add the zeros to the end of the number.)

Thirty-eight and 42 are 2 above and 2 below. Two squared equals 4.

1,600 − 4 = 1,596   ANSWER

That's all there is to it.

Let's try another:

67 × 73 =

If we multiply 67 × 73, we see that each number is 3 away from 70; 67 is 3 lower and 73 is 3 higher. The answer will be 70 squared less 3 squared.

$70^2 = 4,900$

$3^2 = 9$

4,900 − 9 = 4,891   ANSWER

Try these problems for yourself:

a) 27 × 33 =   c) 122 × 118 =

b) 46 × 54 =   d) 9 × 11 =

The answers are:

a) 891   b) 2,484   c) 14,396   d) 99

(The last problem was included as a gift.)

Multiplying Numbers Near 50

Here is another shortcut related to the method for squaring numbers we learned in Chapter Ten.

To multiply two numbers near 50, add the numbers in the circles, halve the answer and add to 25. That will give the number of hundreds in the answer. Then multiply the numbers in the circles. Add this to the answer.

Let's try an example:

$$+④ \quad +⑧$$
$$⑤⓪ \quad 54 \times 58 =$$

Add the digits in the circles. Four plus 8 is 12. Half of 12 is 6. Add 6 to 25.

$$25 + 6 = 31$$

That is the number of hundreds in the answer. (Multiply 31 by 100 to get 3,100.) Multiply the circled digits:

$$4 \times 8 = 32$$

The answer is 3,132.

What if one units digit is odd and the other is even? Let's try it and see.

$$53 \times 54 =$$
$$3 + 4 = 7$$

Half of 7 is 3 ½.

$$25 + 3 ½ = 28 ½.$$

Multiply 28½ × 100 = 28½ hundred, or 2,850. Multiply the circled digits:

$$3 \times 4 = 12$$
$$2,850 + 12 = 2,862 \quad \text{ANSWER}$$

These are easy. Let's try one more:

$$+②+⑬$$
$$⑤⓪ \quad 52 \times 63 =$$

Add the numbers in the circles:

$$2 + 13 = 15$$

Half of 15 is 7½.

$$25 + 7½ = 32½$$

Our working total is 3,250.

Now multiply the numbers in the circles.

$$2 \times 13 = 26$$
$$3,250 + 26 = 3,276 \quad \text{ANSWER}$$

Now try these for yourself:

a) $52 \times 56 =$ c) $53 \times 59 =$

b) $61 \times 57 =$ d) $54 \times 62 =$

The answers are:

a) 2,912   b) 3,477   c) 3,127   d) 3,348

What if the numbers are just below 50?

Let's try an example:

⑤⓪ $46 \times 48 =$
  –④ –②
$4 + 2 = 6$

Half of 6 is 3. Instead of adding this to 25, we subtract it. This is because our multipliers are lower than 50, rather than higher.

$$25 - 3 = 22$$

Our subtotal is 2,200. Multiply the numbers in the circles and add them to our subtotal.

$$4 \times 2 = 8$$
$$2,200 + 8 = 2,208 \quad \text{ANSWER}$$

Let's try another:

⑤⓪ $47 \times 44 =$
  –③ –⑥

Add the numbers in the circles:

$$3 + 6 = 9$$

Half of 9 is 4½. Take 4½ from 25. (Take 5 and add ½.)

$$25 - 4½ = 20½$$
$$20½ \times 100 = 2{,}050$$

Multiply the numbers in the circles:

$$3 \times 6 = 18$$
$$2{,}050 + 18 = 2{,}068$$

These problems can easily be calculated in your head. Try these for yourself:

a) $49 \times 48 =$ b) $46 \times 47 =$

The answers are:

a) 2,352   b) 2,162

Subtraction From Numbers Ending in Zeros

The simple rule for subtracting any number from a number where all the digits are zeros except the first digit is to subtract the final digit of the subtrahend (the number we are subtracting) from 10, then each successive digit from nine. Subtract 1 from the first digit of the minuend (the number we are subtracting from).

Example:

```
  3 0 0 0 0 0    (the minuend)
 - 2 5 7 1 3     (the subtrahend)
  2 7 4 2 8 7    ANSWER
```

We subtract 3 from 10 to get 7. Subtract the other digits from 9.

One from 9 is 8, 7 from 9 is 2, 5 from 9 is 4, 2 from 9 is 7. Subtract 1 from the first digit to get 2.

This is what we do if the number we are subtracting has fewer digits than the one we are subtracting from:

```
  20000000
- 0052316
  19947684
```

We simply add zeros to the number we are subtracting to make up the extra digits.

Because we don't have to worry about carrying or borrowing numbers, we can perform the subtraction from the left or from the right. Calculating from left to right can be very impressive.

This is an easy way to subtract from 100 or 1,000. For example:

$$1,000 - 257 = 743$$

You should be able to call the answer at a glance from left to right. You would say "Seven four three" as you subtract each digit from 9, working from the right. Of course, you would subtract the final digit from 10. And the first digit is easy, as 1 minus 1 is 0.

As you work with numbers you will develop many strategies of your own. At one time in Australia there was a sales tax of 27½ percent. I asked a man who had to make the calculation constantly how he did it. This was in the days before electronic calculators.

He told me that 25 percent plus 2½ percent makes 27½ percent. Twenty-five percent is a quarter of the amount being taxed. If you add one tenth of the quarter you get 27½ percent.

So, if you had to find 27½ percent sales tax on an item costing 80¢, you would first find a quarter of 80. Twenty is 25 percent of 80 and one tenth of 20 is 2. Thus, 27½ percent of an 80¢ item is 22¢.

He and his colleagues developed this simple method to save themselves work. That is how most discoveries are made.

If you are prepared to look for shortcuts and try things differently from how you have been taught, you may make some great discoveries.

Chapter Twenty

Adding and Subtracting Fractions

Fractions are easy. We work with fractions all the time. If you tell the time, chances are you are using fractions (half past, a quarter to, a quarter past, etc.). When you eat a quarter of chicken or talk about football or basketball (half time, second half, etc.), you are dealing with fractions.

We even add and subtract fractions, often without realizing it or thinking about it. We know that two quarters make a half. Half time during a basketball game is at the end of the second quarter.

To calculate that half of 6 is 3, you have also done a calculation involving fractions.

In this chapter, we are going to learn an easy method to add and subtract fractions.

This is a fraction:

$$\frac{1}{2}$$ (the numerator) (the denominator)

The top number is called the numerator and the bottom number is called the denominator.

The bottom number, the denominator, tells you how many parts something is divided into. For instance, a football game is divided into four parts, or quarters.

The top number, the numerator, tells you how many of the parts we are working with. It may be three quarters of a cake or one eighth of a pizza.

$\frac{1}{2}$ is another way of saying 1 divided by 2. $\frac{6}{3}$ means 6 divided by 3 and is another way of expressing 2.

We often have to add, subtract, multiply or divide parts of things. That is another way of saying we often have to add, subtract, multiply or divide fractions.

Here is how we add and subtract fractions.

Addition

Adding fractions is easy. To add $\frac{1}{4}$ + $\frac{2}{3}$ you multiply crossways and then multiply the two denominators.

Let's try an example:

$$\frac{1}{4} \diagup + \diagdown \frac{2}{3} \quad = \quad \frac{3 + 8}{12}$$

We multiply crossways:

$1 \times 3 = 3$

$4 \times 2 = 8$

We add these numbers to get the numerator of the answer.

$3 + 8 = 11$

We multiply the denominators, $4 \times 3 = 12$ to get the denominator of the answer.

The answer is $\frac{11}{12}$. Easy.

Let's try another:

$$\frac{2}{3} + \frac{1}{5} =$$

We multiply crossways.

$$\frac{2}{3} \underset{\times}{+} \frac{1}{5} \; =$$

$2 \times 5 = 10$

$3 \times 1 = 3$

We add the answers for the numerator of the answer.

$10 + 3 = 13$

Multiply the denominators for the denominator of the answer.

$3 \times 5 = 15$

The completed problem looks like this:

$$\frac{2}{3} \underset{\times}{+} \frac{1}{5} \; = \; \frac{10 + 3}{15} \; = \; \frac{13}{15} \quad \text{ANSWER}$$

Here is another example:

$$\frac{2}{3} \underset{\times}{+} \frac{1}{6} \; =$$

Multiply crossways.

$2 \times 6 = 12$

$3 \times 1 = 3$

We add the totals for the numerator, 15. Now multiply the denominators.

$3 \times 6 = 18$

This is the denominator of the answer.

$$\frac{12 + 3}{3 \times 6} \; = \; \frac{15}{18} \quad \text{ANSWER}$$

There is still one more step. Can the answer be simplified?

If both the numerator and denominator are even, we can divide them by 2 to get a simpler answer. For example, $\frac{4}{8}$ could be simplified to $\frac{2}{4}$ and even further $\frac{1}{2}$.

In the above answer of $\frac{15}{18}$, the terms can't be divided by 2, but 15 and 18 are evenly divisible by 3. ($15 \div 3 = 5$, $18 \div 3 = 6$.)

Our final answer is $\frac{5}{6}$.

Each time you calculate fractions, you should give the answer in its simplest form. Check to see if numerator and denominator are both divisible by 2, 3 or 5, or any other number. If so, divide them to give the answer in its simplest form.

For instance, $\frac{21}{28}$ would be $\frac{3}{4}$ (both 21 and 28 are divisible by 7).

Try these for yourself:

a) $\dfrac{1}{4} + \dfrac{1}{3} =$ 　　　　　　c) $\dfrac{3}{4} + \dfrac{1}{5} =$

b) $\dfrac{2}{5} + \dfrac{1}{4} =$ 　　　　　　d) $\dfrac{1}{4} + \dfrac{3}{5} =$

How did you do?

The answers are:

a) $\dfrac{7}{12}$ 　　b) $\dfrac{13}{20}$ 　　c) $\dfrac{19}{20}$ 　　d) $\dfrac{17}{20}$

Another Shortcut

Here is a shortcut. If the numerators are both 1, we add the denominators to get the numerator of the answer (top number), and we multiply the denominators to get the denominator of the answer.

Look at this example:

$$\frac{1}{4} \times \frac{1}{5} = \frac{4 + 5}{4 \times 5} = \frac{9}{20}$$

This method allows you to add and subtract fractions without having to calculate the lowest common denominator and often allows you to "see" the answer at a glance.

You should "see" that:

$$\frac{1}{3} + \frac{1}{4} = \frac{7}{12}$$

and that:

$$\frac{1}{3} - \frac{1}{4} = \frac{1}{12}$$

If you want to add three fractions, add the first two, then add the answer to the third.

For example:

$$\frac{1}{3} + \frac{1}{2} + \frac{2}{5} =$$

First: $\dfrac{1}{3} + \dfrac{1}{2} = \dfrac{3+2}{3\times2} = \dfrac{5}{6}$

Next: $\dfrac{5}{6} + \dfrac{2}{5} = \dfrac{25+12}{6\times5} = \dfrac{37}{30}$

The numerator of the answer, 37, is higher than the denominator so we subtract 30 from 37 (or divide 37 by 30) to get our answer:

$$1\frac{7}{30}$$

Thirty divides once into 37 with 7 remainder.

Subtraction

A similar method is used for subtraction.

Example:

$$\frac{2}{3} - \frac{1}{4} = \frac{8-3}{12} = \frac{5}{12}$$

Again, multiply crossways, to get $2 \times 4 = 8$ and $1 \times 3 = 3$ for the numerator values. Then multiply the denominators for the value of the denominator of the answer.

Try these for yourself:

a) $\dfrac{1}{2} - \dfrac{1}{3} =$ c) $\dfrac{2}{3} - \dfrac{2}{7} =$

b) $\dfrac{3}{4} - \dfrac{1}{7} =$ d) $\dfrac{4}{5} - \dfrac{2}{7} =$

Fractions are easy when you know how. The answers are:

a) $\dfrac{1}{6}$ b) $\dfrac{17}{28}$ c) $\dfrac{8}{21}$ d) $\dfrac{18}{35}$

Chapter Twenty-One

Multiplying and Dividing Fractions

If you add fractions, you end up with an answer that is bigger than any of the individual values. If you subtract fractions, the number you are working with becomes smaller, as you would expect.

Multiplying and dividing with fractions is quite different than with whole numbers and this seems to complicate the subject for many people. Division usually reduces the size of a number, but dividing by a fraction actually makes the number larger. Multiplication by a fraction reduces the size of the number. It seems like a backwards world where nothing makes sense.

When I was a teenager I played football for a junior team. In the third quarter, we had a tradition of eating oranges. We would each get a quarter of an orange. There were 20 players on my football team (including reserves); how many oranges were necessary to feed all 20? Remember, we each received a quarter of an orange.

One orange would feed 4 of us, so 5 oranges would feed 20. An orange divided into quarters makes 4 pieces. Five oranges divided into quarters makes 20 pieces. Five oranges divided into halves would only feed 10 players. Dividing oranges increases the number of orange pieces available.

How about multiplication? If a quarter of the players were injured after a game, how many players would that be? A quarter of 20 is 5.

Why didn't we divide to find out how many players? We could have divided by 4 to find the answer. Dividing by 4 is the same as multiplying by a quarter.

⇨ What answer do you get by multiplying 6 by 10? 60.

⇨ What answer do you get by multiplying 6 by 8? 48.

⇨ What answer do you get by multiplying 6 by 5? 30.

⇨ What answer do you get by multiplying 6 by 2? 12.

⇨ What answer do you get by multiplying 6 by 1? 6.

⇨ What answer do you get by multiplying 6 by $\frac{1}{2}$? 3.

⇨ What answer do you get by multiplying 6 by $\frac{1}{3}$? 2.

This makes sense. The smaller the number you multiply by, the smaller the answer.

So, saying "Half of 6" is the same as saying "Multiply 6 by a half." We know that saying "Half of 6" is the same as saying "6 divided by 2".

Let's go back to our definition for multiplication. Three times 7 means three sevens added together. You would add 7 plus 7 plus 7.

Two times 10 would be 10 plus 10.

What about $1\frac{1}{2}$ times 10?

That would be 10 plus a half of 10. We multiply 10 times $1\frac{1}{2}$ to get 15.

Ten times $\frac{1}{2}$ would simply be half of 10, which is 5.

Multiplying Fractions

You probably know the answer to the following problem without doing any calculation.

$$\frac{1}{2} \times \frac{1}{4} =$$

Let's examine how we would calculate the answer.

We multiply the numbers on the top, the numerators, to get the numerator of the answer.

$$1 \times 1 = 1$$

We multiply the numbers on the bottom, the denominators, to get the denominator of the answer.

$$2 \times 4 = 8$$

The answer is $\frac{1}{8}$.

That's the explanation. Who said fractions were difficult? They are easy. Let's try another.

$$\frac{1}{3} \times \frac{1}{4} =$$

(What we are saying is, "What is a quarter of a third or a third of a quarter?")

$$1 \times 1 = 1 \text{ (the numerator)}$$
$$3 \times 4 = 12 \text{ (the denominator)}$$
$$\frac{1}{12} \quad \text{ANSWER}$$

Let's do one more:

$$\frac{2}{3} \times \frac{1}{2} =$$

Multiply the numerators:

$$2 \times 1 = 2$$

Multiply the denominators:

$$3 \times 2 = 6$$

The answer is $\frac{2}{6}$, which reduces to $\frac{1}{3}$.

Again, we can see this is so because, how many thirds did we have to begin with? Two. We had to work out what is half of two-thirds. Half of 2 is 1, so the answer is one-third.

Try these for yourself.

a) $\dfrac{1}{2} \times \dfrac{1}{3} =$ c) $\dfrac{2}{3} \times \dfrac{2}{5} =$

b) $\dfrac{1}{2} \times \dfrac{1}{5} =$ d) $\dfrac{1}{13} \times \dfrac{1}{14} =$

To calculate the final problem, you would use the method for calculating numbers in the teens in your head.

The answers are:

a) $\dfrac{1}{6}$ b) $\dfrac{1}{10}$ c) $\dfrac{4}{15}$ d) $\dfrac{1}{182}$

How would you multiply $1\frac{1}{2}$ by $3\frac{1}{4}$?

Firstly, you would change them from mixed numbers to improper fractions. A mixed number is one that contains both whole numbers and fractions. Improper fractions are fractions the numerator of which is larger than the denominator.

To change $1\frac{1}{2}$ to an improper fraction, multiply the number (1) by the denominator, (2) to get an answer of 2, and then add the numerator to give a new numerator of 3. The answer is $\frac{3}{2}$. (One and a half is the same as three halves.)

An easy way to understand the procedure is to look at the 1, then the half. How many halves are there in 1? That's easy, 2. Plus one more half for the fraction for a total of 3 halves.

Let's try it for $3\frac{1}{4}$. Multiply the number (3) by the denominator to get 12 quarters, plus one more quarter gives an answer of $\frac{13}{4}$. Now we can write the problem as:

$$\dfrac{3}{2} \times \dfrac{13}{4} =$$

Multiply the numerators: $3 \times 13 = 39$. That is the numerator of the answer. Multiply the denominators: $2 \times 4 = 8$.

The answer is $\frac{39}{8}$.

How do we change the answer back to a mixed number?

We divide 8 into 39. Eight divides into 39 four times ($8 \times 4 = 32$) with 7 remainder.

That gives us: $\frac{39}{8} = 4\frac{7}{8}$.

What if you want to multiply a whole number by a fraction?

Let's try 7 times $\frac{3}{4}$. Another way of expressing this is to say, "What is three quarters of 7?" Three quarters of 8 is 6, so the answer will be slightly less than 6.

Express 7 as a fraction, which would be $\frac{7}{1}$.

$$\frac{7}{1} \times \frac{3}{4} =$$

$$7 \times 3 = 21$$

$$1 \times 4 = 4$$

$$\frac{21}{4} \quad \text{ANSWER}$$

To express this as a mixed number, divide the 21 by 4, which gives us $5\frac{1}{4}$. (Four divides into 21 five times with 1 remainder.)

Dividing Fractions

To find the half of a number you divide by 2. For example, half of 6 is 3. You would write it like this:

$$\frac{6}{1} \times \frac{1}{2} = 3$$

Or you could do one of the following:

$$6 \div 2 = 3$$

$$\frac{6}{1} \div \frac{2}{1} = 3$$

The rule is this:

> **To divide by a fraction, you turn it upside down and multiply.**

$$6 \div \frac{1}{4} = 6 \times \frac{4}{1} = 24$$

Another way to express this would be to ask, "How many quarters can you get from 6 oranges?" You divide 6 oranges into quarters and you get 24 pieces, enough to feed a football team (with reserve players), the coach and trainer, and still have some pieces left over.

Divide two cakes into sixths ($\frac{1}{6}$) and you would have 12 pieces. You could give 12 people one piece each.

Therefore, 2 divided by $\frac{1}{6}$ = 12.

The calculation would look like this:

$$2 \div \frac{1}{6} = \frac{2}{1} \times \frac{6}{1} = \frac{12}{1}$$

Try these division problems:

a) $\frac{1}{3} \div \frac{3}{4} =$ c) $\frac{2}{7} \div \frac{4}{5} =$

b) $\frac{7}{8} \div \frac{2}{3} =$

The answers are:

a) $\frac{4}{9}$ b) $\frac{21}{16}$ c) $\frac{5}{14}$

The third problem would be calculated:

$$\frac{2}{7} \div \frac{4}{5} = \frac{2}{7} \times \frac{5}{4} =$$

$$2 \times 5 = 10 \text{ (numerator)}$$

$$7 \times 4 = 28 \text{ (denominator)}$$

The answer is $\frac{10}{28}$. Both the numerator and the denominator are even numbers so we can divide them by 2 to get our answer of $\frac{5}{14}$.

Chapter Twenty-Two

Direct Multiplication

A simple way to multiply numbers that don't suggest any easy reference number is by the direct method. This is the usual method employed by people who are lightning calculators.

For example:

36 × 72 =

Here is how you would set the problem out in your head:

You would calculate from left to right, beginning with 70 times 30. You would multiply 7 × 3 and multiply the answer by 100. (In practice, you can multiply 7 × 3 and then add the two zeros to the answer.)

7 × 3 = 21

21 × 100 = 2,100

This is our first working total. Now multiply crossways, 7 × 6 and 3 × 2 and add the answers.

7 × 6 = 42

3 × 2 = 6

42 + 6 = 48

Multiply this answer by 10 and add to our working total:

48 × 10 = 480

2,100 + 480 = 2,580

If you say, "Twenty-one hundred plus four hundred is twenty-five hundred, plus the eighty is twenty-five hundred and eighty," you will have no trouble doing this mentally.

Then multiply the units digits. Six times 2 is 12. Add 12 to our running total to get an answer of 2,592.

2,580 + 12 = 2,592   ANSWER

By commencing the calculation from the left you get an approximate answer with your first step. Each further calculation refines your answer.

And all of this can be calculated entirely in your head.

Let's try another:

34 × 73 =

We set it out like this:

We multiply 7 × 3 = 21, plus two zeros (one zero for each digit after the digits we are multiplying) gives a running total of 2,100.

Now we multiply crossways:

(3 × 3) + (7 × 4) =

9 + 28 = 37

We add one zero for the cross multiplied numbers because there was one digit after each multiplication. That gives us 370.

We would say, "Twenty-one hundred plus three hundred and seventy is twenty-four hundred and seventy." We add, "Twenty-one hundred plus three hundred is twenty-four hundred, plus seventy is twenty-four hundred and seventy."

Our running total is 2,470.

Now we multiply the units digits.

$$4 \times 3 = 12$$
$$2,470 + 12 = 2,482 \quad \text{ANSWER}$$

Try these for yourself.

a) 42 × 74 =   c) 27 × 81 =

b) 37 × 64 =   d) 34 × 72 =

Are you impressed by how easy it is to calculate these problems in your head?

The answers are:

a) 3,108   b) 2,368   c) 2,187   d) 2,448

This method is extremely useful when no easy shortcut is apparent.

Multiplication by Single-Digit Numbers

Direct multiplication by a single-digit number is easy.

To multiply 43 by 6, you would multiply 40 by 6 and then add 3 sixes. To multiply 40 by 6, you multiply 6 × 4 and add a zero to the number.

$$6 \times 4 = 24$$
$$24 \times 10 = 240$$
$$3 \times 6 = 18$$
$$240 + 18 = 258$$

That was easy. It was easier than using reference numbers and our general multiplication formula.

How about 6 × 17?

Six times 10 is 60, plus 6 sevens are 42. This gives an answer of 102.

Let's say we want to calculate 6^3. That means 6 × 6 × 6, three sixes multiplied together.

We multiply the first two sixes:

$$6 \times 6 = 36$$

Now we multiply the answer, 36, by 6. To do this we multiply 30 by 6, then 6 by 6, and add the answers:

$$6 \times 30 = 180$$

Plus $6 \times 6 = 36$ gives us:

$$180 + 36 = 216$$

To add 180 plus 36, I would add 20 from the 36 to make 200, and then add the other 16 to get the final answer of 216.

Direct multiplication by a single-digit number is easy and almost automatic.

Try these for yourself:

a) $7 \times 13 =$ d) $9 \times 26 =$

b) $8 \times 23 =$ e) $6 \times 124 =$

c) $6 \times 42 =$ f) $8 \times 206 =$

They were not difficult. The answers are:

a) 91   b) 184   c) 252

d) 234   e) 744   f) 1,648

Here's a tip for e), multiply 6×120 and then add 6×4.

Did you get them all correct? Most people feel these calculations look too difficult and won't try them. If you can give an immediate answer you will appear to be highly intelligent and gifted at mathematics.

Multiplying Larger Numbers

Let's try 123×45:

```
  1  2  3        1  2  3        1  2  3        1  2  3
   \              \ |            \ /            |
    \              \|             X             |
  ×  4  5        ×  4  5        ×  4  5        ×  4  5
  ──────────     ──────────     ──────────     ──────────
  4  0  0  0     5  3  0  0     5  5  2  0     5  5  3  5
```

To find the answer we would multiply 1 times 4. Then 1 times 5 plus 2 times 4. Then 2 times 5 plus 3 times 4. Then, finally, 3 times 5.

You can see how the pattern works.

$$100 \times 40 = 4,000$$

(I would say, "Four times 100 is 400, times 10 because we are multiplying by 40, 4,000.")

One hundred times 5 is 500, plus 20 times 40, 800, is 1,300.

$$4,000 + 1,300 = 5,300$$

Twenty times 5 is 100, plus 3 times 40, 120, is 220.

$$5,300 + 220 = 5,520$$

Five times 3 is 15.

$$5,520 + 15 = 5,535 \quad \text{ANSWER}$$

To arrive at our answer, our running totals were 4,000, 5,300, 5,520 and 5,535.

Standard multiplication would first give you the units digit of 5. Although you can use the direct multiplication from left to right or from right to left, beginning with the higher digits gives us an immediate answer close to the actual answer.

Here is another way to visualize our calculation:

1 2 3 × 4 5 =

One hundred times 40 equals 4,000. Working total is 4,000.

1 2 3 × 4 5 =

Five times 100 is 500 plus 20 times 40 equals 800, gives 1,300. Working total is 5,300.

1 2 3 × 4 5 =

Five times 20 is 100 plus 3 times 40 is 120, which gives 220. Working total is 5,520.

1 2 3 × 4 5 =

Three times 5 equals 15. Final total of 5,535.

To multiply 321 × 427 with pencil and paper, you need only write the answer.

3 2 1 × 4 2 7 = 120,000

Write a zero for each digit after the digits you are multiplying.

3 2 1 × 4 2 7 = 134,000

3 2 1 × 4 2 7 = 136,900

3 2 1 × 4 2 7 = 137,060

3 2 1 × 4 2 7 = 137,067   ANSWER

You can keep track of your place with your fingers.

Try these problems for yourself. At first, try them with pencil and paper. Then, just write the answer or calculate them mentally.

a) 123 × 345 =

b) 204 × 436 =

c) 623 × 316 =

d) 724 × 315 =

The answers are:

a) 42,435 b) 88,944 c) 196,868 d) 228,060

Combining Methods

You can combine direct multiplication with our method of using a reference number. We try to choose simple reference numbers of 10, 20, 50, and 100. If you want to use 30 or 70 as reference numbers, then you can use them combined with the direct multiplication method.

If you wanted to multiply 68 × 68 for instance, you could use 70 as a reference number.

(70)   68 × 68 =
     –② –②

Subtract diagonally.

$$68 - 2 = 66$$

To find the subtotal we multiply 66 by the reference number, 70. Using direct multiplication:

$$70 \times 66 =$$
$$60 \times 70 = 4{,}200$$
$$6 \times 70 = 420$$
$$4{,}200 + 420 = 4{,}620$$

Now multiply the numbers in the circles and add the answer to our subtotal:

$$2 \times 2 = 4$$
$$4{,}620 + 4 = 4{,}624$$

Actually, you could use a shortcut for this example. You could factor 66 to 6×11.

Now the calculation becomes $7 \times 6 \times 11 \times 10$.

$$7 \times 6 = 42$$
$$42 \times 11 = 462 \quad \text{(elevens short cut)}$$
$$462 \times 10 = 4{,}620$$

Multiplying the numbers in the circles and adding this to our subtotal gives us our answer of 4,624.

Using the direct method of multiplication will make it possible to use any reference number.

Chapter Twenty-Three

Estimating
Answers

There are many times we need to estimate answers. How much will it cost to run my car this year? How much will the repairs cost me? How much bank interest will I pay or be paid? Likewise, when we calculate a currency conversion it is impossible to get an accurate answer. Rates change every day. We don't know what the bank will charge as a commission. At best, our calculation is an estimate.

Once, I was in a supermarket buying some food for my dinner and I suddenly realized I only had a $20 bill in my pocket.

I checked the food in my cart and discovered the food added to approximately $22. From the cart I chose something worth about $3.50, put it back on the shelf, and took my goods to the checkout.

This was in the days before the scanners, so the cashier manually entered the prices into the register.

She told me, "That's twenty-six dollars and forty cents."

I said, "I am sorry, you have made a mistake. It is less than twenty dollars."

The cashier burst into tears and the manager hurried over.

"What appears to be the trouble?" he asked.

She explained, "This man says I have made a mistake with his total."

"How do you figure that?" he asked, looking at me with suspicion.

I explained I only had 20 dollars in my pocket and I had bought accordingly.

We checked the prices against the register receipt and found the cashier had failed to insert a decimal. Had I not done a mental estimate, I would never have noticed the mistake. Later, I discovered it was her first day on the job and I felt a little guilty.

There are many times we need to make quick estimates in mathematics. We don't need an accurate answer, we just need an approximate answer. Estimating is a skill you can easily learn.

How do we estimate what we will have to pay in the supermarket? Round off all prices to the nearest dollar or half dollar. Some you will round upwards, some downwards. That makes it easy to get an approximate answer that will tell you if you have sufficient in your pocket to pay for the purchases. Try it when you go to the supermarket and see how accurately you can estimate the total.

Practical Examples

How many people are there in the auditorium?

We look around and it is almost full. Count the number of rows. We count 16. How many people in a row? We count 20 chairs, but the average row seems to have 14 people. The numbers vary from row to row, but 14 seems to be the average.

We multiply 14 by 16 to get 224. Even though our answer is an exact number, it is still only an estimate. We would say, "Just over two hundred people."

How much is a vacation going to cost?

Allow so much per night for hotel bills, so much for gas and your rental car, so much for meals, and so much for incidentals such as souvenirs. Adding these together will give you an approximate price. Vacation experience tells us to add 50 percent to the answer because everything costs more than you think.

What is the cost of a used car?

The answer is, whatever you pay for the car, plus at least $1,000 for repairs you may find it needs after you buy it. I tell people, budget for the repairs. If they aren't necessary, you have come out ahead. It is a bonus. If you do have to pay, you don't feel ripped off.

What is the estimated cost of 227 articles at $485 each?

If we don't know the final price we will be charged, we can estimate it and hopefully negotiate the price down. About how much will they cost? We can round the numbers off to 200 times 500 to get an answer of $100,000. (You will notice that I rounded one number upwards and the other downwards to minimize any error.)

We could also estimate another way, and multiply 230 × 480 using the methods in this book:

$$
\begin{array}{l}
\textcircled{\scriptsize 50} \quad 23 \times 48 = \\
-\textcircled{\scriptsize 27} \ -\textcircled{\scriptsize 2}
\end{array}
$$

$$21 \times (100 \div 2) = 1{,}050$$
$$-27 \times -2 = 54$$
$$1{,}050 + 54 = 1{,}104$$

Our answer is approximately $110,000. Both answers are valid estimates. Hopefully we could then negotiate the price down to about $105,000.

The simple method for estimation is to round off all numbers to the number of significant digits we need in the answer. Make up the rest of the digits with zeros.

Your brother bought 253 used photocopiers for $10,000. How much did he pay per copier?

To calculate an exact answer, you would divide 10,000 by 253. If you don't have a calculator with you, you can give a close estimate by rounding off the numbers.

To estimate the answer, you would divide 10,000 by 250. Two hundred and fifty is a quarter of one thousand, so you can divide by 1,000 and multiply by 4.

Ten thousand divided by 1,000 is 10. (What you are asking is, how many thousands make up ten thousand?)

Ten times 4 is 40. The answer is just under $40 per copier. This is probably close enough for your requirements. The actual answer is $39.53 per copier.

When estimating answers for addition, subtraction, multiplication and division, you simply round off the numbers. If you round one number upwards, round the other number downwards.

In these days of cheap electronic calculators, why do we need to estimate when our calculator will give an exact answer?

Firstly, in the above examples, you are not sure of the figures; you can only put approximate figures into your calculator. Secondly, calculators have actually increased the need for mental estimation. I will explain why.

I once asked a class to make the following calculation.

Gas sells at $1.30 per gallon. You put 18 gallons of gas in your car. How much do you pay?

A student gave an answer of several million dollars. I asked him if it was the correct answer. The student replied yes, because he obtained it with his calculator.

I then asked him if $1.30 was a reasonable price for a gallon of gas. The student again replied yes. I asked him if his father's car fuel tank could hold 18 gallons. He said yes again, and told me the size of the fuel tank.

Finally, I asked him if his father had ever paid $3,000,000 for a tank of petrol. Then it sank in. The calculator gave an answer but it wasn't correct. He must have pressed a wrong button. The answer was actually $23.40.

Many people accept a calculator answer without question. We need to be able to estimate answers so that we can see if we have made a mistake using our calculator.

Chapter Twenty-Four

Using What You Have Learned

We encounter mathematics every day. Every time we make a purchase, every time we hear or see the time, every time we get in our car we make mathematical calculations. How much time do I have? Six forty is the same as twenty to seven. How much do I give the cashier? Do I have enough for this purchase? How much will I have to budget to pay for my house, my mobile phone? Do I have enough time to drive to town? These are all mathematical calculations. In this chapter we will see how a basic knowledge of math can help us in our everyday lives.

Traveling Abroad

When people travel to another country, some enjoy the differences in tradition, culture and language; others are terrified by the same differences.

I enjoy the diversity. If everything were the same, I might as well stay home. When visiting a foreign country I like to think in terms of that country's currency, measurements, temperatures, etc. Often, though, we may want to change the measurements in the country we are visiting to measurements our friends and family at home can understand. Also, we often need to convert currencies to our own so that we can tell if we are making a good purchase. This all requires mathematics.

Temperature Conversion

When traveling to a country that uses degrees Fahrenheit rather than degrees Celsius, or vice versa, it is helpful if we can easily convert from one to the other. If you hear the weather forecast, you want to know whether you need your overcoat or your summer clothing. The formula to change from Fahrenheit to Celsius is to subtract 32 degrees and multiply the answer by 5 and then divide by 9. That is not difficult using the methods taught in this book, but we might not need an exact equivalent. Here is a simple formula:

To change Fahrenheit to Celsius, subtract 30 degrees and then halve.

To change Celsius to Fahrenheit, double and then add 30.

The answer you will get using this formula will not be exact, but it's close enough for practical purposes. For instance, if you are told the temperature will be 8°C, double it and add 30. Double 8 is 16, plus 30 makes 46°F. The actual temperature would be 46.4°F. That's close enough for practical purposes.

What if you are visiting a country which measures the temperature in Fahrenheit, and you are used to Celsius? If you are told the temperature will be 72°F, subtract 30 (42) and then halve. The answer is 21°C. The actual answer is about 22°C. We are out by one degree, but we have a fair idea of what the temperature will be.

Easy temperature conversion formulas:

$$(°C \times 2) + 30 = °F$$
$$(°F - 30) \div 2 = °C$$

Exact temperature conversion formulas:

$$°C \times \frac{9}{5} + 32 = °F$$
$$°F - 32 \times \frac{5}{9} = °C$$

Try converting the following temperatures in your head using each formula.

Change 80°F to °C

Using the simple formula, subtract 30 from 80 to get 50, then halve for the temperature in Celsius. 80°F = 25°C.

Using the exact formula:

$$80 - 32 \times \frac{5}{9} =$$

$$80 - 32 = 48$$

$$48 \times 5 = 240$$

$$\frac{240}{9} = 26.67°C$$

Let's try another:

Change 10°C to °F

Using the exact formula:

$$10 \times \frac{9}{5} + 32 = 90$$

$$\frac{90}{5} = 18$$

$$18 + 32 = 50°F$$

Or, using the simple formula, double 10°C to get 20, then add 30 to get your answer of 50°F.

In the first example our simple formula gave an answer that was close; in the second the answer was exact.

If you feel you might forget when you have to add or subtract the 30 degrees, or when you halve or double the temperatures, memorize an equivalent temperature and then see how you would make the calculation.

For instance, if you knew that 100°F is the same as 37 or 38°C, how would you change 100 to get 37? One hundred minus 30 is 70. Half of 70 is 35. Close enough. If we want to change back, double 35 to get 70, then add 30 to get 100. We have checked the formula in both directions.

Another generally well-known set of equivalents is the fact that water freezes at 0°C and 32°F. Take 30 from 30°F to get 0°C.

It is good to have a median value memorized. Ten degrees Celsius is equal to 50°F. Work up or down from it. 20°C = 68°F. (A change of 10 degrees Celsius equals a change of 18 degrees Fahrenheit.)

Let's try an example:

> Convert 15°C to °F

The actual formula:

$$15 \times \frac{9}{5} + 32 =$$

$$15 \times 9 = 135$$

$$\frac{135}{5} = 27$$

$$27 + 32 = 59°F$$

Using our easy formula:

> 15 x 2 = 30   30 + 30 = 60°F—1 degree error

One more example:

> Convert 20°C to °F
>
> $$20 \times \frac{9}{5} + 32 =$$
>
> $$\frac{180}{5} = 36$$
>
> $$36 + 32 = 68°F$$

Using the easy formula:

> 20 x 2 = 40   40 + 30 = 70°F—2 degrees error

This is pretty close. It does give an idea of temperatures when you are traveling.

Time and Distances

Another common conversion required when traveling is the conversion factor from inches to centimeters. A foot rule (another name for the common ruler) is 30 cm. Thirty divided by 12 (the number of inches on the ruler) gives us approximately 2½. That gives us approximately 2.5 centimeters to the inch.

For time differences it is a good idea to memorize a time. Work out a favorable time to phone home or your work and memorize the time equivalent.

For example, I spend a lot of time traveling between Melbourne and Vancouver. Often I need to know the time in both cities. I simply memorize that 12 noon summer time in Melbourne is 5 p.m. the previous day in Vancouver. To convert 2 p.m. Melbourne summer time to Vancouver time I add 2 hours to 5 p.m. and get 7 p.m. When it is 2 p.m. Vancouver time it is 9 a.m. Melbourne time. Here, rather than adding and subtracting 19 hours to or from the local time, I work backwards and forwards from my memorized time.

Money Exchange

When I visited the United States recently, the Australian dollar was worth around 65¢ US. To calculate how much a U.S. item was worth in Australian dollars, I divided its price by 0.65. Alternatively, I could have doubled the price and then divided by 1.3, or divided the price itself by 1.3 and simply doubled it. Also, look at the reciprocal. One U.S. dollar is worth about $1.50 Australian. That makes it easy to change. Forty dollars US will be worth half as much again in Australian dollars, about $60 Australian. The exchange rates have changed since, but it is easy to calculate similar equivalent values when you travel.

Speeds and Distances

Memorize the fact that 100 km is about 60 miles. Multiply or divide by 0.6 to make the conversion.

Remember that 60 m/hr is a mile a minute. Traveling at 100 km/hr, it will take 30 minutes to travel 30 miles.

I also keep in mind times for traveling at 100 km per hour. It will take two and a half hours to travel 250 km. To travel 25 km takes a quarter of an hour. So, if I still have 175 km to drive to reach my destination, I know it will take me about one and three-quarter hours to get there.

How fast is 50 km per hour in m/hr? If 100 km equals 60 miles, 50 km must be 30 miles. Fifty km/hr is about 30 m/hr.

Pounds to Kilograms

One kilogram is equal to 2.2 pounds. To change kilograms to pounds you multiply by 2.2. To change pounds to kilograms you divide by 2.2, which is the same as 11×0.2.

How heavy is someone who weighs 65 kg?

$$65 \times 0.2 = 13$$
$$13 \times 11 = 143 \text{ lbs.}$$

Sports Statistics

While you are watching a sports game, either at the field or on television, calculate the statistics for yourself. What is the number of runs per at bats, what is a player's on base percentage, and what is a pitcher's earned run average? Almost any game has its own statistics and calculations. You can become more involved and improve your math ability at the same time.

Estimating Distances

When I drive in Vancouver, British Columbia, in Canada, I know that the streets and avenues are numbered eight to the mile. If I am just crossing 16th and I have to drive to 86th, then I know how far I have to drive and can calculate approximately how long it will take me to arrive. I have 70 streets to go. Seventy divided by 8 is almost 9. I have almost 9 miles to drive or, if I want to be more exact, 8¾ miles to go.

Miscellaneous Hints

Measure the diameter of coins and then use them to measure when you don't have a ruler or tape measure. Measure the size of your thumb joint. Measure your hand span. Measure your average stride. Measure the length of your foot and your shoes. What is the length of your arm? What is the distance between the tips of your fingers when your arms are outstretched? What is the size of legal or US letter paper size? What is the size of A4 paper?

Use all of these measurements to estimate sizes.

Walk a kilometer or a mile and count your steps. Then divide the number of steps taken into the distance to get the length of an average stride. Time how long it takes to walk a mile or a kilometer. You can then calculate your average speed and estimate how long it will take to walk a certain distance or calculate how far you have walked.

Count audiences when you attend a movie or a concert. Count the number of seats by multiplying the number of seats in a row by the number of rows. What is the seating capacity of the theater?

Do you know the circumference of the earth? It is approximately 24,000 miles or 40,000 km. At the equator, the earth moves about 1,000 miles on its axis every hour. In other words, there are about 1,000 miles between hourly time zones along the equator. Because the earth has a circumference of about 24,000 miles, that makes it easy to divide into hours of the day. If you want the distance in kilometers, 1,000 miles is about 1,600 kilometers.

Apply the Strategies

Make use of the strategies you have learned in this book, not only for school or work, but also for play. Use the calculations when you are driving or traveling. Use them to calculate how your favorite team is doing. Use the strategies when you are shopping. And, of course, use them on the job and at school. You will develop your skills, sharpen your brain and make better decisions.

Afterword

Recently, I was teaching a fifth-grade class. I had just explained how to multiply numbers in the teens when I noticed a girl experimenting with 109 times 109. She drew the circles above, using 100 as the reference number, and obtained an answer of 11,881. She asked me if this was correct. She wasn't asking if she had made an error in her calculations; she was asking if the method was valid for her calculation. I assured her she was correct on both counts.

When I teach a class as a visiting teacher, this is a common reaction. Children want to experiment. This is probably the most pleasing result of my teaching as far as I am concerned. Children begin to think like mathematicians. Also, because they see results for their efforts, they are happy to put in even more effort. Children often ask their teachers if they can do math for the rest of the day.

Making your own mathematical discoveries is exhilarating. The methods taught in this book develop creative thinking and problem-solving skills. They lay the groundwork for original thought, lateral thinking and the ability to work outside set guidelines. These strategies instill an excellent foundation of basic mathematical principles, and a knowledge of simple number facts. The methods are easy to master and applicable to everyday situations. Use what you have learned. Play with the strategies. Experiment with them.

Students

Practice the strategies taught in this book and you will gain a reputation for being a mathematical genius. You will breeze through your math courses and find math exciting and enjoyable.

Teachers

Teach these methods in the classroom and math classes will be a pleasure for you and your students. Your students will be successful, which will make you a successful teacher. Also, because your students are achieving, you will find classroom behavior will improve and your students will be motivated. Everyone benefits.

Parents

Teach your children these methods and they will excel at mathematics. They will not only be faster than the other students; they can check their work and correct any mistakes before anyone else sees them. The methods will boost your children's confidence, not only in math, but in their own intelligence as well. Your children will be treated differently because people equate math ability with intelligence. They will probably improve in other subject areas as well.

Many children have a low self-esteem and believe they are "dumb." They assume they don't have a "mathematical mind" because they can't handle basic math problems. Yet, after learning my methods, many parents have written to say their children now love math. Their children are excited at their success and the recognition they are receiving. They have discovered they don't have an inferior mind. It wasn't the quality of their brain—it was their problem-solving techniques that were at fault.

I have enjoyed writing *Speed Mathematics* and experimenting with the strategies. I have deliberately tried to keep the book non-technical so everyone can understand the information it contains.

I have placed practice sheets on my website which can be downloaded. The website also contains details of other material that I offer. If you would like to comment on this book, want details of new books and other learning materials, or if you have any questions, please email me at bhandley@speedmathematics.com or visit my web page at www.speedmathematics.com.

Appendix A

Frequently
Asked Questions

Q. *My child is already at the top of the class in mathematics. Won't your methods make my child bored? What do children do if they have finished their work in a quarter of the time it takes the other children?*

Q. *If I use your methods I will finish my work in less time and I won't have anything to do. I will get bored.*

A. Students who use these methods love to experiment. Firstly, they will finish their assignments in a fraction of the time it takes the rest of the class. Then they can check their work for mistakes by casting out nines and elevens. They can experiment with alternative methods of calculation to see which method is easiest. Actually, students who learn these methods find math exciting.

Q. *What about understanding? If you use this method to learn your tables, it doesn't teach you why 6 times 7 is 42.*

A. No, it doesn't. Nor does any other method of teaching the tables. Chanting and reciting tables don't teach the student why six sevens are 42. My strategy is to teach an effective and easy method to determine the answer.

And, although the reason why the method works is not immediately apparent, the method itself can be easily explained by the time the child reaches fourth grade. (See the explanation as to why the formula works in Appendix D). Any fourth-grade child who has worked through this book should be able to understand the explanation.

What 6 sevens means should be taught as well as how to calculate the answer. Working by rules is not sufficient. You need an understanding of mathematics. Students who learn by these methods generally have a better understanding and are less likely to work blindly by rules.

Q. *If the school is teaching different methods, won't this confuse my child?*

A. No. These methods complement what students learn in school. Higher achieving students use different methods than the low achievers. Sometimes that can cause confusion among the teachers, but it seldom affects the students. Most of the methods are invisible. That is, the difference is what takes place in the student's mind. If the student chooses to say nothing about his or her different methods, no one will ever know.

Q. *My children's teachers demand the children show all of their work for their calculations. If the calculations are in one's head or different, how can they do this?*

A. Students can give what their teachers want. If the student is taking an exam, naturally the student will give everything he or she believes the teacher or examiner wants.

In an ordinary class, if the teacher were to ask to show the work for 13 times 14, it should be sufficient for the student to say, "I know my tables up to the 20 times table. I don't have to work them out." If the teacher challenges the child, the child should be able to give quick answers to any multiplication in the teens. The child can also do a lightning casting out nines calculation to check the answer immediately. The teacher will be impressed, not annoyed.

Q. *Your method is not always the easiest method. Why should I use your method when there is an easier alternative?*

A. By all means, use the method you find easiest. I am suggesting some easy methods—you should consider the methods you know and use the one you find easiest.

For instance, if you wanted to multiply 8 times 16, you could use the circles and 10 as your reference number. I wouldn't. I would probably multiply 8 times 10 and add 8 times 6 (80 + 48 = 128). Or, I might multiply 8 times 8 equals 64 and double the answer.

I feel an important part of my teaching strategy is to give students a choice of methods. I had a student come up to me in the schoolyard and say, "I am sorry, Mr. Handley, I am not using your methods anymore."

"Why not?" I asked.

"I know my tables now and I just call them up from memory."

Did I consider this a bad thing? Not at all. The student had just told me she had memorized the tables beyond 15 times 15.

Students who learn these methods are less likely to solve problems according to rules or a formula. They are more likely to be innovative in their approach.

Q. *Why do you teach students how to calculate these problems? What do we have calculators for?*

A. A calculator can't think for you. Students have a much better appreciation of the properties of numbers if they learn these methods. They understand the principles involved in calculations rather than working by rules if they master the methods in this book.

When I am asked this question in the classroom, I ask students to take out their calculators and to enter the following problem as I call it.

I tell them to press the buttons as I call them.

$$2 + 3 \times 4 =$$

Some calculators give an answer of 20. Others give an answer of 14. Fourteen is the correct answer.

Why the two different answers? Some calculators know BODMAS, or the order of mathematical functions. You multiply before you add or subtract. The problem is really "two plus three fours." Three fours are 12, added to the 2 is 14.

A calculator can't think for you. A calculator isn't much help if you don't understand mathematics. The best familiarity with numbers and mathematical principles comes from learning the kinds of strategies taught in this book.

Q. *Are you for or against calculators?*

A. Calculators are useful tools. They can save a lot of time and effort. There are many times during the day when I use a calculator.

Students sometimes ask me, "How would you multiply sixteen million, three hundred and forty-nine thousand, six hundred and eighty-nine by four million, eight hundred and sixty-two thousand, one hundred and ninety-four?" I tell them my first step would be to take out my calculator. Often they seem shocked at my answer. I often use a calculator. I add up columns of numbers with a calculator. I double-check my answer because I know it is possible to make a mistake.

I also mentally estimate the answer to see if the calculator's answer "makes sense." The answer should be close to my estimation.

When scientific calculators first became available, I purchased a cheap calculator. I discovered I didn't understand all of the functions and made sure I learned what the functions did. My calculator was the cause of my learning some areas of statistics that I wasn't aware of.

I have often wondered what some of the mathematical geniuses of the past would have done with the scientific calculators of today. I am sure they would have made good use of them and maybe achieved much more than they did.

Q. *Did you make up these methods yourself?*

A. Yes, I made up many of the strategies. I made up the ideas with the circles and reference numbers, but I was taught factoring for long multiplication and division by my third- and fourth-grade teachers, Miss Clark and Mrs. O'Connor. Miss Clark taught the method for subtraction, and long multiplication by factors. Mrs. O'Connor taught the method of long division by factors. I have picked up many other strategies along the way.

When I was in elementary school I recognized the easy method for adding and subtracting fractions but—can you believe it—I was too shy to speak up in class and tell anyone.

Q. *I worked out some of these strategies for myself and I have always done better at math than my classmates. It isn't fair that you have taught them what I do and made them as good as I am. I deserve some advantage and recognition for working them out for myself.*

A. An American high school student asked this question. I think he has a point. My answer is that when we teach mathematics, we should teach the best strategies and give the best explanations possible. We shouldn't leave it to the "more intelligent" students to work them out for themselves. Why not enable everyone to succeed?

Q. *Teaching these kinds of strategies will make poor students into good students. Many of these students will lose their friends as they become more intelligent. Aren't you causing social problems?*

A. This question was asked after I had given an after-dinner talk at an international service club. I am still not sure if the questioner was serious, but the reactions of the other members suggested that he was.

I would rather deal with the problems caused by improving someone's intelligence than those caused by a lack of intelligence and a lack of success.

Q. *I am a new teacher, recently out of college. Will I get into trouble if I teach these strategies in the classroom? What will happen if my fourth-grade students are doing sixth-grade work by the end of the year?*

A. If there is an easy method for teaching a subject and a difficult method, who could teach the difficult method in good conscience? If, in teaching the three and four times tables, students also master the six, seven, eight and nine times tables, is this bad? You are teaching what you are required to teach, but not in the way you were taught to teach it.

These methods fall within curriculum guidelines because they teach what students are required to learn, and then some. My ninth-grade math teacher, Harry Forecast, taught us ninth-grade math with tenth- and eleventh-grade math thrown in for good measure. I loved learning math under his teaching. I couldn't wait to get home to try logarithms for myself. He taught many shortcuts as part of his teaching strategy. I felt like Sherlock Holmes solving a difficult case when I used some of his methods to solve problems in algebra.

To answer the question, fifth- and sixth-grade teachers should be pleased the students are ahead and should use the opportunity to stretch the students even further. I believe that when these methods are generally taught in schools this will happen. I trust that this book will bring this about.

Q. *I am a new teacher too. I have always been terrified of math. What if I teach your methods and get stuck myself? What if students ask questions I can't answer? Maybe it would be safer to just teach the way everybody else does. Wouldn't I be taking a risk teaching your way?*

A. Sure, you would. But you can minimize the risk. The methods are not difficult. Begin slowly. Teach the strategy for calculating the tables up to 10 times 10. Let them play with the strategy for a day or two. Then, give them numbers in the nineties to calculate. They are doing similar calculations but with much greater satisfaction. While they are calculating these problems, they are not only memorizing their lower tables (from multiplying the numbers in the circles) but also the higher tables, the combinations of numbers that add to 10, and also basic subtraction. It is then an easy step to teach subtraction from numbers in the teens. $14 - 8 = 4 + 2 = 6$. (See Chapter Nine.)

Then, as you teach multiplying numbers in the teens you are introducing positive and negative numbers. You don't have to give a full explanation. Tell them the subject will be covered later.

As you teach these methods, you will find your own ability with numbers will improve. Your confidence will grow. Tell the students you are learning the methods along with them. This will tend to make the subject and the methods more exciting to the children, and they will be more forgiving if you make a mistake.

Estimating Cube Roots

It is not often we have to estimate cube roots, but there are times when we need to know the size of a sphere or a box. For example, if we were repaving our driveway, we may need to calculate a cubic value of liquid, sand or gravel. For most people, there is only one way to calculate cube roots, and that is with a calculator. Even then, they could only use a scientific calculator.

The cube root of 27 is 3 because $3 \times 3 \times 3 = 27$. You multiply three threes together to cube three. Three cubed is 27. The cube root of 27 is 3. This would be written as $\sqrt[3]{27}$. The 3 in front of the root sign tells you it is the cube root. (Convention dictates that square roots can have the 2 before the sign, but leaving out the number in front the root sign also refers to the square root.)

I will show you an easy method for estimating cube roots just as we calculated square roots by estimation. And, if you have a cheap four-function calculator, you can calculate the answer using our simple formula.

Firstly, you have to memorize the cubes of the numbers 1 to 10. Here are the values:

1^3	=	1	4^3	=	64
2^3	=	8	5^3	=	125
3^3	=	27	6^3	=	216
7^3	=	343	9^3	=	729
8^3	=	512	10^3	=	1,000

The cubes of the numbers 1 to 5 are easily learned because they are easily calculated if you forget.

Our first step in finding the square root of a number was to split the number into digit pairs, whereas:

> **To find the *cube* root we split the digits into *threes.* The number of three-digit groups gives us the number of digits in the answer.**

We then estimate the cube root of the first digit triplet. This is where you need to memorize the above chart.

⇨   If the first three-digit group falls between 1 and 7, the first digit of the answer is 1.

⇨   If it falls between 8 and 26, the first digit is 2.

⇨   If it falls between 27 and 63, the first digit is 3.

You should have the idea. The estimate of the first triplet gives the first digit of the answer. Designate a zero for each of the other triplets. This is your first estimate of the answer.

Let's take an example of 250:

$$\sqrt[3]{250} =$$

Two hundred and fifty is above 6 cubed, 216, but below 7 cubed, 343. This tells us the answer will be between 6 and 7.

We divide by our estimate, 6, and we divide twice.

$$250 \div 6 = 41.67$$

Divide our answer, 41.67, by 6 again:

$$41.67 \div 6 = 6.94$$

The difference between our estimate, 6 and our final answer, 6.94, is 0.94. Divide this by 3 and add it to our estimate.

$$0.94 \div 3 = 0.31$$

Adding this to 6 gives 6.31.

$$_3\sqrt{250} = 6.31$$

This will always be slightly high so we round off downwards to 6.3. A calculator gives an answer of 6.2996. We didn't round off sufficiently, but our answer is correct to two significant figures. And the real advantage is that the above calculation can easily be done mentally.

Another way of describing the final step is to take the average of the three numbers. To find the average, we calculate 6 + 6 + 6.94 and divide by 3.

$$6 + 6 + 6.94 = 18.94$$
$$18.94 \div 3 = 6.31$$

I find it much easier to divide the difference by 3.

Using a ten-digit four-function calculator, I used 6.31 as a second estimate and repeated the calculation. I got a final answer of 6.2996053, whereas my scientific calculator gave an answer of 6.299605249—so our answer was accurate to seven digits.

Try these calculations for yourself:

a) $_3\sqrt{230}$ c) $_3\sqrt{8162}$

b) $_3\sqrt{540}$ d) $_3\sqrt{30{,}000}$

The answers are:

a) 6.127   b) 8.1457   c) 20.134   d) 31.07

Using the above method, your answers should have been very close. You might like to calculate your percentage error.

There is another shortcut you could use for c) and d). The first estimates for the answers would have been 20 and 30. You could have done one single division by 20^2 and 30^2. That would have meant dividing by 400 and 900. You would have moved the decimal two places to the left and divided by 4 and 9.

In a similar fashion to our method of estimating square roots, if the number we are calculating is just below a cube we can use the number above as the estimate. Then divide twice by the estimate and subtract a third of the difference from the estimate. Again, there is a similar shortcut for this calculation to make it a reasonable option.

Let's take an example of the cube root of 320.

$$_3\sqrt{320} =$$

Six cubed is 216 and 7 cubed is 343. Seven is definitely a closer estimate.

$$320 \div 7 = 45.71$$

We divide by 7 again:

$$45.7 \div 7 = 6.53$$

Subtract 6.53 from 7:

$$7 - 6.53 = 0.47$$

We now need to calculate a third of the difference.

$$0.47 \div 3 = 0.157$$

Subtract a third of the difference (0.157) from our estimate, 7.

$$7 - 0.157 = 6.843$$

Round off to 6.84 and we have an accurate answer.

$$_3\sqrt{320} = 6.84$$

The actual answer is 6.8399.

Now for the shortcut. What we really did was to average the three values, 7 and 7 and 6.53. We could add these values and divide by 3.

$$7 + 7 + 6.53 = 20.53$$
$$20.53 \div 3 = 6.843$$

Dividing 3 into 20 we get an answer of 6 with 2 remainder which we carried to the 0.53 to give us 2.53.

$$2.53 \div 3 = 0.843$$

Just as we know there is a 1 carried for square roots, we know a 2 is carried for cube roots. Instead of subtracting a third of the difference from the higher estimate, we use the lower number (in this case 6), carry 2 and divide by 3 to get our end answer.

Why do we carry 2 when we are calculating cube roots? Because, like the example above, when we average the three values we add the two numbers which are 1 above the answer. So the total must be three times the estimate plus 2.

Let's try another to illustrate.

$$_3\sqrt{700} =$$

Our estimate is 9, which we know is high (9 cubed is 729).

Divide 700 by 9 twice.

> 700 ÷ 9 = 77.77 (rounded off downwards)
> 77.77 ÷ 9 = 8.64

The first digit of the answer is 8. For the rest of the answer, put 2 in front of the decimal and divide by 3.

> 2.64 ÷ 3 = 0.88

Our answer is 8.88. This answer is accurate to three digits.

Let's do one more:

$$_3\sqrt{7,531} =$$

We split the digits into groups of three.

We now have:

$$_3\sqrt{7\ 531} =$$
$$* \quad *$$

Estimate the cube root of the first three-digit group, 7. Seven is close to 8 which is 2^3 so we will use 2 as our first estimate. There are two three-digit groups so there are two digits in the answer. We assign zero as the second digit, making our estimate 20.

Divide 7,531 twice by 20. To divide by 20, we divide by 10 and then by 2.

$$7,531 \div 20 = 376.55$$
$$376.55 \div 20 = 18.8275$$

Or, instead of dividing by 20 twice, we could have divided by 20 squared, or 400.

$$7,531 \div 100 = 75.31$$
$$75.31 \div 4 = 18.8275$$

We know the first digit, 1, is correct. This gives us the tens digit. Our subtotal is 10.

We place a 2 in front of the rest of the number to get 28.8275.

$$28.8275 \div 3 = 9.609$$
$$10 + 9.609 = 19.609$$

Rounding off we get an answer of 19.6. This is accurate to three digits. Actually the answer is 19.60127, which means our answer is extremely close.

Try these for yourself and then check your work with the explanations below:

a) $\sqrt[3]{115} =$ b) $\sqrt[3]{500} =$

For a) we take 5 as our estimate.

We divide 115 by 5 to get 23. (Divide by 10 and double.) We divide 23 by 5 to get 4.6. (Divide by 10 and double.) Four is the first digit of our answer.

We carry 2 to put before the 6 to get 26. Then we divide by 3.

$$2.6 \div 3 = 0.8667$$

We round off to 4.86. This is accurate to three digits.

For b), we take 8 as our first estimate.

$$500 \div 8 = 62.5$$
$$62.5 \div 8 = 7.8125$$

To average the answers we carry 2 to the 8, giving 2.8125.

Dividing 2.8125 by 3, we get 0.9375. We add this to the 7 to get 7.9375.

The actual answer given by a calculator is 7.93700526. Our answer is extremely accurate for a mental calculation or an easy pencil and paper calculation. Also, we know our answers are high. Had we rounded off downwards we would have been right on.

This is an effective and a very impressive method for finding cube roots. The average person wouldn't dream of trying to calculate the answer with pencil and paper. This method will allow mental calculation.

Appendix C

Checks for Divisibility

It is easy to check to see if a number is evenly divisible by another number without doing a full division.

Here are some basic rules:

(1) All numbers are divisible by 1.

(2) All even numbers are divisible by 2. (If the final digit of a number is even, the number is divisible by 2.)

(3) If a number is evenly divisible by 3, the sum of the digits is divisible by 3. For instance, 12 is divisible by 3 as 1 + 2 = 3, which is divisible by 3.

(4) If the last two digits of a number are divisible by 4, the number is divisible by 4. For example, 116 is divisible by 4. We can see this because 16 is 4 times 4.

(5) If a number ends with a zero or a 5, the number is divisible by 5.

(6) If a number is even and the sum of its digits is divisible by 3, it is divisible by 6.

(7) * (See the note at the end of this list.)

⑧   If the last three digits of a number are divisible by 8, the number is divisible by 8. For example, 1,128 is divisible by 8 because 128 equals 8 × 16.

⑨   If the digit sum of a number adds to 9 or is evenly divisible by 9, the number is divisible by 9.

⑩   If a number ends with a zero, the number is divisible by 10.

⑪   If the difference between the sum of the evenly placed digits and the oddly placed digits in a number is zero or a multiple of 11, the number is evenly divisible by 11.

⑫   If the digit sum of a number is divisible by 3 and the last two digits are divisible by 4, the number is divisible by 12.

⑬   *

⑰   *

⑲   *

⑳   If the tens digit of a number is even and the units digit is zero the number is divisible by 20.

㉑   If a number is divisible by 7 and the sum of the digits is divisible by 3 the number is divisible by 21.

㉓   *

㉙   *

* There is a simple method, using check multipliers, for checking divisibility that can be used for these numbers and higher numbers. The traditional methods are cumbersome and require pen and paper for the calculation. These checks can be done mentally.

Using Check Multipliers

To check divisibility by 7, we use 5 as a check multiplier. We multiply the units digit of the number we are checking by our check multiplier.

We add this answer to the number after removing the units digit. If this answer is evenly divisible by 7, the whole number is evenly divisible by 7.

For example, is 91 evenly divisible by 7?

Our check multiplier is 5. Multiply the units digit of 91 (1) by 5 to get an answer of 5. Add 5 to the 9 to get an answer of 14, which is 2 times 7. So 91 is divisible by 7.

And, is 133 evenly divisible by 7?

Multiply the units digit of 133 (3) by our check multiplier, 5, to get an answer of 15. Add the answer, 15, to 13 to get 28 (4 × 7). Thus, we find 133 is divisible by 7.

Let's try another. Is 152 evenly divisible by 7?

Multiply 2 by 5 equals 10. Add 10 to 15 to get 25. Twenty-five is not evenly divisible by 7, so 152 is not evenly divisible by 7.

One final example. Is 1,638 evenly divisible by 7?

$$5 \times 8 = 40$$
$$163 + 40 = 203$$

Because we may not know off the top of our heads whether 203 is evenly divisible by 7, we repeat the procedure:

$$5 \times 3 = 15$$
$$20 + 15 = 35$$

Thirty-five is divisible by 7 (5 × 7 = 35). Therefore, 1,638 is evenly divisible by 7.

How to Determine the Check Multiplier

The method for determining the check multiplier is as follows:

For a positive check multiplier, multiply the number to get an answer with a units digit of 9. We use the tens digit of the number that is 1 more than this answer as our check multiplier.

For instance, if we want to check for divisibility by 7, multiply 7 by 7 to get 49. Forty-nine is one below 50. Our check multiplier for 7 is then 5.

To check for divisibility by 13, we multiply it until we find a units digit of 9:

13 × 3 = 39

Thirty-nine is 1 below 40. We use 4 as our check multiplier for 13.

With 19, we don't need to multiply it by anything, it already ends in 9. Nineteen is 1 below 20 so we use 2 as the check multiplier.

To check divisibility by 23, we note that the units digit is 3, and 3 threes are 9. Multiplying 23 by 3 we get 69. This is 1 below 70, so we use 7 as the check multiplier.

Why Does It Work?

If you want to check whether a number is evenly divisible by another number, adding the number or a multiple of the number won't affect the divisibility.

When we checked if 91 was evenly divisible by 7, we actually added 49 (7 × 7) to 91 to get a total of 140. Dropping the zero at the end of 140 doesn't alter the outcome.

To check if 112 is evenly divisible by 7, we multiply 2 times 5 to get 10.

Then, 11 + 10 = 21, which is 3 × 7.

Adding 100 to 110 is the same as adding 98 (2 × 49 or 7 × 7 × 2) to 112.

98 + 112 = 210

210 = 3 × 7 × 10

Let's examine the check multiplier for 13 more closely. Firstly, what is the check multiplier for 13?

3 × 13 = 39

Thirty-nine is one below 40 so we use 4 (the tens digit of 40) as our check multiplier. To test for divisibility by 13, multiply the units digit by four and add the tens digit.

For example, is 78 evenly divisible by 13?

The units digit is 8:

$$8 \times 4 = 32$$
$$32 + 7 \text{ (the tens digit of 78)} = 39 \quad (3 \times 13)$$

Because 39 is three times 13, then 78 is divisible by 13.

If you weren't sure of 39, you could repeat the process.

$$9 \times 4 = 36$$
$$36 + 3 = 39$$

Because we end up with the same number, we know the number is evenly divisible.

Let's try another. Is 351 divisible by 13?

The units digit is 1:

$$1 \times 4 = 4$$
$$4 + 35 = 39 \quad (39 = 3 \times 13)$$

Yes, 351 is divisible by 13.

How about 3,289? Is it evenly divisible by 13?

The units digit is 9:

$$9 \times 4 = 36$$
$$328 + 36 = 364$$

We don't know if 364 is divisible by 13 so we can use the test over again.

The units digit of 364 is 4.

$$4 \times 4 = 16$$
$$36 + 16 = 52, \text{ which is } 13 \times 4$$

If we didn't know that 13 times 4 is 52, we could try once again.

The units digit of 52 is 2.

$$2 \times 4 = 8$$
$$5 + 8 = 13$$

Now we know for certain that 3,289 is evenly divisible by 13.

What about the other numbers on our list at the start of the chapter?

To check for divisibility by 17, use 12; by 19, use 2; by 23, use 7; by 29 use 3 as a check multiplier. These numbers can be found by using the rule for determining check multipliers.

For example, is 578 evenly divisible by 17?

Most of us don't feel inclined to even try a division by 17. We would probably just use a calculator.

We know that 12 is our check multiplier for 17. Multiply the 8 of 578 by 12.

$$12 \times 8 = 96$$
$$96 + 57 = 153$$

It is still not obvious if 153 is a multiple of 17. Try it again.

$$3 \times 12 = 36$$
$$36 + 15 = 51$$

If you are not sure if 51 is a multiple of 17, keep repeating the process.

$$1 \times 12 = 12$$
$$12 + 5 = 17$$

Obviously, 17 is evenly divisible by 17. Therefore, 578 is a multiple of 17.

Note: I teach an alternative method for checking divisibility by 17 later in this appendix.

Let's try an example the other way around. Seven times 13 equals 91, so 91 should test as divisible by both numbers.

Test for 7:

$$1 \times 5 = 5$$
$$5 + 9 = 14$$

Fourteen is 2×7, so 91 is evenly divisible by 7.

Test for 13:

$$1 \times 4 = 4$$
$$4 + 9 = 13$$

Therefore, 91 is evenly divisible by 13.

Try these for yourself:

    a) Is 266 evenly divisible by 19?

    b) Is 259 evenly divisible by 7?

    c) Is 377 evenly divisible by 13?

    d) Is 377 evenly divisible by 29?

The answer is yes to all of the above questions.

Negative Check Multipliers

There is also a negative multiplier for divisibility checks:

For a negative check multiplier, multiply the number to get an answer with a units digit of 1. We use the tens digit of this answer as our check multiplier.

The negative check multiplier for 17 would be 5, because $3 \times 17 = 51$. Let's repeat an earlier test we did with positive check multipliers:

Is 578 evenly divisible by 17?

Our negative check multiplier is minus 5 (–5).

$$-5 \times 8 = -40$$

We subtract our answer from the rest of the number.

$$57 - 40 = 17$$

In one step we found 578 is evenly divisible by 17.

Let's try another. Is 918 divisible by 27?

Firstly, we need to determine our negative check multiplier for 27. Three times 27 is 81. Our negative check multiplier is minus 8.

We multiply the units digit of 918 by our check multiplier, minus 8.

$$-8 \times 8 = -64$$
$$91 - 64 = 27$$

Hence, 918 is evenly divisible by 27.

One more example. Is 135 evenly divisible by 27?

The check multiplier is minus 8.

$$5 \times -8 = -40$$
$$13 - 40 = -27$$

The answer is minus 27, which tells us 135 is evenly divisible by 27. (To get our answer, all we had to do was subtract 13 from 40 and place a negative sign in front of it.)

Try these for yourself:

a) Is 136 evenly divisible by 17?

b) Is 595 evenly divisible by 17?

c) Is 1426 evenly divisible by 31?

d) Is 756 evenly divisible by 27?

The answer is that they are all evenly divisible by the numbers given. These are very easy checks.

Positive or Negative Check Multipliers?

You would normally use negative check multipliers for numbers ending in 7 or 1. Let's look at the possible units digits of the numbers we might want to check.

If the number ends in 1, we already have our negative check multiplier. It is the number in front of the 1. For example, 31 would have a negative check multiplier of 3.

If the number ends in an even digit, we would halve it and use either rule for half the number.

If the number ends in 3, we could multiply by 3 to get a positive check multiplier ending in 9.

If the number ends in 5, we would divide the number by 5 and use the rule for the answer.

If the number ends in 7, we would multiply it by 3 to get a negative check multiplier ending in 1.

If the number ends in 9, we would add 1 and use the number in front of the zero as a positive check multiplier.

Why Our Methods Work

Multiplication With Circles

Why does it work?

Firstly, let me give the "easy" explanation.

Let's multiply 99 × 85.

The standard shortcut would go like this.

Ninety-nine is almost 100, so multiply by 100 and subtract 85.

$$85 \times 100 = 8,500$$

Now we have to subtract 85. What is the easy way to subtract 85?

Subtract 100 and add 15.

$$8,500 - 100 = 8,400$$
$$8,400 + 15 = 8,415$$

Doesn't that look suspiciously like our method with the circles?

For the same problem, 99 × 85, with the circles we take one from 85 to get 84 and multiply by 100 to get 8,400. Then, because we only subtracted *one* hundred, we added *one* 15 to the answer.

For 98 × 85 we could multiply by 100 and subtract 2 times 85.

$$85 \times 100 = 8,500$$

Subtract 2 times 85 from our answer. What is the easy way to do that?

Instead of adding 85 plus 85 and subtracting from 8,500, let's take 100 twice and add 15 each time for the correction. Taking the 200 first gives us 8,300.

Instead of adding 15 and then another 15, just say 2 times 15 is 30 and simply add 30. The answer is 8,330.

We can extend this reasoning to multiplying numbers below 10.

$$9 \times 8 =$$

Say 10 times 8 is 80, then subtract 8 to get 72. Working with circles we get:

$$\overset{\textstyle(10)}{\underset{\textstyle -(1)-(2)}{9 \times 8 \, = \, 72}}$$

Let's try one more example: $7 \times 8 =$

$$\overset{\textstyle(10)}{\underset{\textstyle -(3)-(2)}{7 \times 8 \, =}}$$

Here, if we multiply 10 times 7 and then subtract 2 times 7 from that answer, we can see a correlation between methods. Ten times 7 is 70. The easy way to subtract 2 times 7 is to subtract 2 times 10 and then give back 2 times 3.

That is what I would call the "easy" explanation of why multiplication with circles works. Even elementary school children can follow this reasoning—especially after mastering the methods taught in this book.

Algebraic Explanation

Now for the explanation using algebra.

Let's consider the problem:

$$13 \times 14 =$$

$$\overset{\textstyle +(3) \; +(4)}{\underset{}{(10) \quad 13 \times 14 \, =}}$$

Let "a" equal the reference number, in this case, 10; "b" and "c" are the units digits or the values in the circles, in this case, 3 and 4.

The calculation can be written as:

$$(a + b) \times (a + c) =$$

Multiplying $(a + b) \times (a + c)$ we get:

$$a^2 + ab + ac + bc$$

The first three terms are then divisible by "a," so we can multiply them by "a" outside a parenthesis.

$$a (a + b + c) + bc$$

Substituting numbers for $13 \times 14 =$ we get,

$$(10 + 3) \times (10 + 4) =$$
$$10 (10 + 3 + 4) + (3 \times 4) =$$
$$10 \times 17 + 12 =$$
$$170 + 12 = 182$$

With this formula, "b" and "c" can represent either positive or negative values, depending on the position of the circles. In the calculation for 7×8, "b" and "c" would be negative values.

The above formula is used to square numbers near 50 and numbers ending in 5.

It is not necessary for a child to understand this formula to be able to use it effectively. (See Appendix A: Frequently Asked Questions.)

Using Two Reference Numbers

We can express the formula as:

$$(a + b) \times (xa + c)$$

We let "a" equal the reference number, "b" and "c" are the numbers in the circles, and "x" is the multiplication factor.

We can multiply this out for:

$$xa^2 + xab + ac + bc$$

The first two terms are divisible by a, so we can reduce this to:

a(xa + xb + c) + bc

Let's try this with an actual calculation.

13 × 41 =

Our base reference is 10 and our second reference is 40, 4 × 10. The numbers in the circles are 3 and 1. Our calculation should look like this:

③ ①

(10 × 4) 13 × 41 =

a = 10 (base reference number)
b = 3   (the circled number above 13)
c = 1   (the circled number above 41)
x = 4   (the multiplication factor)

Substituting numbers for our formula we get:

a(xa + xb + c) + bc
10(4 × 10 + 4 × 3 + 1) + (3 × 1)
= 10 (40 + 12 + 1) + (3 × 1)
= 10 × 53 + 3
= 530 + 3 = 533   ANSWER

The full calculation would look like this:

⑫

③ ①

(10 × 4) 13 × 41 = 530

                        +3
                    ─────
                     533

Formulas for Squaring Numbers Ending in 1 and 9
1. Squaring numbers ending in 1

To square 31, we square 30 to get 900.

Then we double 30 to get 60 and add it to our previous total.

$$900 + 60 = 960$$

Then add 1.

$$960 + 1 = 961$$

This is simply cross multiplication or direct multiplication.

To multiply 31 by 31 you could make the same calculation using the algebraic formula.

$$(a + 1)^2 = (a + 1) \times (a + 1)$$
$$(a + 1) \times (a + 1) = a^2 + 2a + 1^2$$

In the case of 31^2, a = 30.

We squared 30 to get 900. We doubled "a" to get 60. We didn't need to square 1 because 1 remains unchanged after squaring.

The benefit of the formula is that it keeps the calculation in an easy sequence and allows for easy mental calculation.

2. Squaring numbers ending in 9

Squaring numbers ending in 9 uses the same formula as those for ending in 1 but with a negative 1.

Example:

$$29^2 =$$

To calculate 29^2, we would round it to 30. We square 30 to get 900. Then we double 30 to get 60 and subtract this from our subtotal.

$$900 - 60 = 840$$

Then we add 1.

$$840 + 1 = 841$$

The standard formula is $(a + 1) \times (a + 1)$. In this case 1 is negative so we can write it:

$$(a - 1) \times (a - 1)$$

Multiplying this out we get:

$$a^2 - 2a + 1$$

This is precisely what we did for 29^2.

Remember, "a" is 30. We squared 30 to get 900. This time we minus 2a (60) from 900 to get 840. Minus 1 squared $(-1)^2$ remains 1 which we added to get our final answer of 841.

This procedure is simpler than standard cross multiplication.

Adding and Subtracting Fractions

This concept is simply based on an observation I made in elementary school. You don't need to find the lowest common denominator to add and subtract fractions.

If you multiply the denominators together the result must be a common denominator. Then, if you wish, you can cancel the denominator to a lower or the lowest common denominator. If you don't cancel to the lowest common denominator, you might make the calculation a little harder, but you will still end up with the correct answer.

To take a simple example:

$$\frac{1}{2} + \frac{1}{4} =$$

Multiply the denominators to get the denominator of our answer, 8. Add the denominators to get our numerator of 6.

Our answer is $\frac{6}{8}$.

We should immediately see that this cancels to $\frac{3}{4}$ because both the numerator and the denominator are divisible by 2. In this case, the lowest common denominator would have been 4.

Either method is a valid means of reaching the answer.

I would introduce the concept of lowest common denominator in the classroom only after children are confident working with adding and subtracting fractions by my method.

Appendix E

Casting Out Nines— Why It Works

Why does casting out nines work? Why do the digits of a number add to the nines remainder?

Here is an explanation.

Nine equals 10 minus 1. For each 10 of a number, you have 1 nine and 1 remainder. If you have a number consisting of 2 tens (20) you have 2 nines and 2 remainder. Thirty would be 3 nines and 3 remainder.

Let's take the number 32. Thirty-two consists of 30, 3 tens, and 2 units, or ones. Finding the nines remainder, we know that 30 has 3 nines and 3 remainder. The 2 ones from 32 are remainder as well, because 9 can't be divided into 2. We carry the 3 remainder from 30, and add it to the 2 remainder from the 2 ones.

$3 + 2 = 5$

Hence, 5 is the nines remainder of 32.

For every 100, 9 divides 10 times with 10 remainder. That 10 remainder divides by 9 again once with 1 remainder. So, for every 100 you have 1 remainder. If you have 300, you have 3 remainder.

Another way to look at the phenomenon is to see that:

$1 \times 9 = 9 \ (10 - 1)$
$11 \times 9 = 99 \ (100 - 1)$

$$111 \times 9 = 999 \ (1{,}000 - 1)$$
$$1{,}111 \times 9 = 9{,}999 \ (10{,}000 - 1)$$

So, the place value signifies the nines remainder for that particular digit.

For example, in the number 32,145, the 3 signifies the ten thousands—for each ten thousand there will be 1 remainder. For 3 ten thousands there will be a remainder of 3. The 2 signifies the thousands. For each thousand there will be a remainder of 1. The same applies to the hundreds and the tens. The units are remainder, unless it is 9 in which case we cancel.

This is a phenomenon peculiar to the number 9. It is very useful for checking answers and divisibility by nine. It can be used not only to divide by nine, but also to illustrate the principle of division.

Appendix F

Squaring Feet
and Inches

When I was in elementary school and we had to find square areas involving feet and inches, the method they taught us was to reduce everything to the same value—in this case, inches—and multiply.

For instance, if we had to find the area of a garden bed or lawn with the dimensions 3 feet 5 inches by 7 feet 1 inch, we would reduce the values to inches, multiply them and then divide by 144 to bring the values to square feet with the remainder as the square inches.

However, there is a much easier method.

We were taught this method in algebra class but its practical uses were never fully explained.

Let's multiply 3 feet 5 inches by 7 feet 1 inch using our method of direct multiplication.

Firstly, we will call the feet values "f." We would write 3 feet 5 inches by 7 feet 1 inch as:

$$(3f + 5) \times (7f + 1)$$

We set it out like this.

$$
\begin{array}{r}
3f + 5 \\
\times\ 7f + 1 \\
\hline
\end{array}
$$

Here we use the direct multiplication method we learned in Chapter Twenty-Two.

Firstly, we multiply 3f times 7f to get 21f². (That's 21 square feet.)

Now multiply crossways:

3f × 1= 3f, plus 7f × 5 = 35f (That's 35 feet inches.)
3f + 35f = 38f

Our answer to date is 21f² + 38f.

Now multiply the inches values.

5 × 1 = 5

Our answer is 21f² + 38f + 5.

This means we have an answer of 21 square feet plus 38 feet inches plus 5 square inches. (Thirty-eight feet inches means 38 areas of one foot by one inch. Twelve of these side by side would be one square foot.) Divide 38f by 12 to get 3 more square feet, which we add to 21 to get 24 square feet.

Multiply the 2 remaining feet inches by 12 to bring them to square inches:

2 × 12 = 24
5 + 24 = 29 square inches

Our answer is 24 square feet and 29 square inches.

This is a much simpler method of calculating the answer. This method can be used to multiply any values where the measurements are not metric.

Try these calculations for yourself:

a) 2 feet 7 inches × 5 feet 2 inches =

b) 3 feet 5 inches × 7 feet 1 inch =

The answers are:

a) 13 ft² 50 in²   b) 24 ft² 29 in²

How did you do? Try the calculations again, this time without pen or paper. You are performing like a genius. That's what makes it fun.

Appendix G

How Do You Get Students to Enjoy Mathematics?

I am often asked, how can I get my children or students to enjoy mathematics? Why don't students enjoy mathematics?

Are mathematical games the answer? By having students compete in games or quizzes? Certainly, I have seen teachers motivate a class with a game or competition that had every student involved and motivated but, if a child is struggling and cannot do the calculations involved, these activities can also be very discouraging.

Firstly, I believe the major reason people in general say they "hate mathematics" is not that they really hate mathematics but that they hate failure. They equate mathematics with failure. Which sports do you enjoy playing? Usually, the sports you are reasonably good at.

People generally equate mathematical ability with intelligence. If we are good at math, we are intelligent; if we do poorly and struggle we are not so smart. Students not only believe this about others, they believe it about themselves. No one likes to feel that he or she is unintelligent, especially in front of a class of friends and peers.

The most certain way to get students to enjoy mathematics is to enable them to succeed. This is the object of my methods, to enable those who have failed in the past to succeed. It is one thing to tell a student, "You can do it." It is quite another to get the student to believe it.

We all want to succeed. I often address a class of students and tell them what they will be doing in 10 minutes' time. I teach them how to do it and they find to their surprise they are actually doing it. Suddenly they are performing like mathematical geniuses. Usually, children become so excited with their achievements they ask if they can do math for the rest of the day. Children come home from school and excitedly tell their parents and family what they can do. They want to show off their new skills. They want to teach their friends who don't know about the methods.

Remove the Risk

I always tell a new group or class of students that I don't mind where they are now, mathematically, shortly they will be performing like geniuses, and I am going to show them how every step of the way.

When I give the first calculations, how to multiply 7 times 8, I tell them they can count on their fingers if they like. If they want, they can take off their shoes and socks and count on their toes—I won't be offended. I tell them they will all know their basic number facts after just a few days—counting on their fingers will only last a short while.

I give the class plenty of easy problems but have them achieve something that they are impressed by and proud of, like 96 times 97. Even if the students don't know their basic number facts at the beginning, they will after just a couple of days as they practice the methods.

Give Plenty of Encouragement

As the children succeed in their efforts, tell them what they are doing is remarkable. Make sure your encouragement is genuine. It is not difficult to find genuine points to praise when students learn the methods in this book. For example:

"Most students in higher grade levels can't do what you are doing."

"Can you solve the problem in your head? Fantastic!"

"Do you know it used to take three weeks to learn what we have done this morning?"

"Did you think 10 minutes ago you would be able to do this?"

Tell the children together and individually that you are proud of them. They are doing very well. They are one of the best classes you have ever taught. But be careful. As soon as you are insincere in your praise, the students will detect it.

Tell Stories to Inspire

Tell the students true stories of mathematicians who have done remarkable things in the past. Stories of lightning calculators, stories of Tesla, Gauss, Newton, Von Neumann; there are many stories to inspire students. Look for books of stories for children or seek them on the internet.

I have had students come to me after a class and ask, "Do you really think I could be an Einstein?"

Find Mentors

If you can find someone who loves mathematics and has a story to tell, invite him or her to visit your school and speak to the students.

Tell the students your own story. You have a story to tell—even if it is how you discovered the methods in this book. We all need heroes to relate to—why not make mathematical heroes for the children?

Give Puzzles

Give easy puzzles for students to play with. Give puzzles to solve of different levels of difficulty. Make sure that everyone can solve at least some of the puzzles. Teach the students methods of problem-solving.

Find puzzle books that not only give puzzles suitable for your class but also give good explanations of how to solve them.

Ask mathematical questions. Bring everyday mathematics into the classroom. Point out to a child each time he or she uses math or needs math skills. Ask questions that require a mathematical answer, like:

"Which is cheaper, how much will it cost?"

"How much further do we have to travel? What speed have we averaged? How long will it take if we continue at the same speed?"

"Which is cheaper, to drive four people in a car or to travel by train? Or to fly?"

"How much gas will we use to drive to _____? How much will it cost?"

"How much will it cost to keep a horse/pony?"

"How many of us are in our class?"

"If we sit three to a table, how many tables do we need?"

"If we each need 10 books, how many for the class?"

"If a third of the books are damaged by water, how many were damaged? How many are left? At $23 per book, how much will it cost to replace them?"

Instead of just giving a list of questions for students to answer, make the questions part of your conversation with the class. Work out problems with the class. Encourage the children to bring their own puzzles to class.

How to Get Children to Believe in Themselves

1. Tell them they can do it.

2. Show them how they can do it.

3. Get them to do it.

4. Do it with them if necessary.

5. Tell them they have done it—they can do it again.

6. Fire their imagination—tell them to imagine themselves succeeding. What would it be like if . . . ? Imagine yourself . . .

7. Tell them success stories. Inspire them.

Appendix H

Solving Problems

1. **Work on the assumption that you *can* solve the problem, and you *will* solve it.**
   Then, at least, you will begin the process.

2. **Simplify the numbers.**
   See what you do to solve a simple problem. (Instead of $47.36, what if it were $100.00, or $1.00?) Simplifying the numbers can often give you the clue you need. Note your method to solve the "obvious" problem and apply it to your "complicated" problem.

3. **Do the problem backwards.**
   Work from the answer backwards and see what you did. It often helps to combine this with method 2.

4. **Go to extremes—millions or zero.**
   Sometimes this will make the method obvious.

5. **Make a diagram.**
   Draw a picture. This may clarify the problem.

6. **Reverse the details.**
   What if it were the other way around?

7. **Start, find something you *can* do.**
   Doing something, even if it seems to have nothing to do with the answer, will often give you the clue you need. Maybe you will find your step was an important part of solving the problem.

8. **Look for analogies.**
   Is this problem similar to anything else you know?

9. **Visualize the problem.**
   Some logic problems are best solved by "seeing" the situation in your mind.

10. **Make no assumptions—go back to the beginning.**
    Question what you know.

11. **Substitute—use different terms. Take out (or add) emotional elements.**
    So for example, what if it were us, China, Iceland, your mother?

12. **What would you do if you *could* solve the problem?**
    You would at least do something—not just sit there. Try something!

13. **Look for trends.**
    Ask yourself, if this increases, does that increase? What is the big picture?

14. **Trial and error.**
    This is often underrated. It is a valid strategy that will often give the clue to the correct method.

15. **Keep an open mind.**
    Don't be too quick to reject strategies or ideas.

16. **Understand the problem.**
    What are they asking? Have I understood the question correctly?

Glossary

Addend:	One of two or more numbers to be added.
Constant:	A number that never varies—it is always the same. For example, π is always 3.14159. . .
Common denominator:	A number into which *denominators* of a group of fractions will evenly divide.
Denominator:	The number which appears below the line of a fraction.
Difference:	The result of subtracting a number by another number. (The answer to a subtraction problem.)
Digit:	Any figure in a number. A number is comprised of digits. For instance, 34 is a two-digit number. (See also *place value*.)
Dividend:	A number which is to be divided into another number.
Divisor:	A number used to divide another number.
Exponent:	A small number which is raised and written after a base number to signify how many times the base number is to be multiplied. 3^2 signifies that two threes are to be multiplied (3 is the base number, 2 is the exponent). 6^4 signifies $6 \times 6 \times 6 \times 6$.
Factor:	A number which can be multiplied by another number or other numbers to give a *product*. The factors of 6 are 2 and 3.

250

Improper fraction: A fraction of which the *numerator* is larger than the *denominator*.

Minuend: A number from which another number is to be subtracted.

Mixed number: A number which contains both a whole number and a fraction.

Multiplicand: A number which is to be multiplied by another number.

Multiplier: A number used to multiply another number.

Number: An entire numerical expression of any combination of digits, such as 10,349 or 12,831.

Numerator: A number which appears above the line of a fraction.

Place value: The value given to a digit because of its position in a number. For example, 34 is a two-digit number. Three (3) is the tens digit so its place value is 3 tens. Four (4) is the units digit so its place value is 4 units.

Product: The result of multiplying two or more numbers. (The answer to a multiplication problem.)

Quotient: The result of dividing a number by another number. (The answer to a division problem.)

Square: A number multiplied by itself. For example, the square of 7 (7^2) is 49.

Square root: A number which, when multiplied by itself, equals a given number. For instance, the square root of 16 ($\sqrt{16}$) is 4.

Subtrahend: A number which is to be subtracted from another number.

Sum: The result of adding two or more numbers. (The answer to an addition calculation.)

Terms for Basic Calculations

23	Addend
+ 14	Addend
37	Sum

654	Minuend
− 142	Subtrahend
512	Difference

123	Multiplicand
× 3	Multiplier
369	Product

385	Dividend
÷ 11	Divisor
35	Quotient

Index